A LITTLE BOOK OF SELF CARE

TRIGGER POINTS

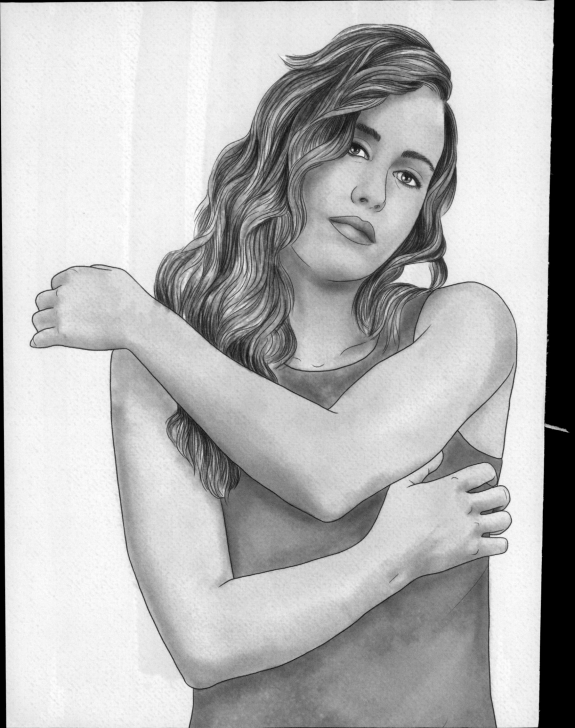

A LITTLE BOOK OF SELF CARE

TRIGGER POINTS

USE THE POWER OF TOUCH
TO LIVE LIFE PAIN-FREE

AMANDA OSWALD

Senior Editors Lesley Malkin, Rona Skene
Senior Designers Mandy Earey,
Collette Sadler
Designers Sadie Thomas, Philippa Nash
Editorial Assistant Kiron Gill
Illustrator Fortuna Todisco
Senior Production Editor Tony Phipps
Production Controller Denitsa Kenanska
Jacket Designer Amy Cox
Jacket Co-ordinator Lucy Philpott
Managing Editor Dawn Henderson
Managing Art Editor Marianne Markham
Art Director Maxine Pedliham
Publishing Director Mary-Clare Jerram
Special Sales Creative Project Manager
Alison Donovan

First published in Great Britain in 2019 by
Dorling Kindersley Limited
DK, One Embassy Gardens,
8 Viaduct Gardens, London, SW11 7BW

DISCLAIMER see page 144

A CIP catalogue record for this book is available
from the British Library.
ISBN: 978-0-2413-8454-1

Printed and bound in China

For the curious
www.dk.com

CONTENTS

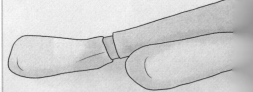

FOREWORD

For many people, modern medicine is literally a lifesaver. However, many of us are increasingly looking for ways to live healthier lives and to self-treat our health conditions. This book is a practical, hands-on guide for anyone interested in how treating your trigger points can bring relief after years of chronic pain or misdiagnosis.

If you've ever felt a "knot" in your shoulder and dug in with your fingers to find both tension and relief, then the chances are you've found a trigger point – a knot of connective tissue that is tender to the touch and that causes pain or other symptoms. The effects of a trigger point may be felt either where the trigger point is located or elsewhere in the body – this is called "referred" pain. It is the possibility of these, sometimes surprising, referred symptoms that mean trigger points can be overlooked medically and sometimes misdiagnosed as a better-known medical condition.

In my clinical work, I see many people experiencing a wide range of chronic pain conditions. Many have found their way to me after years of suffering, rounds of medical treatment, and eventually being told they will just have to live with it. I work with these clients to identify their trigger points as the first step to successfully relieving their pain and treating their condition.

My journey into trigger points started with my own chronic pain and a quest to heal myself. I spent many years working in stressful jobs, alternating long hours with vigorous sports including rugby and long-distance running. Like many people, I dismissed the warning aches and pains that came and went from time to time.

Eventually I developed a prolapsed disc. The pain forced me to stop working and abandon the leisure activities I loved, so I accepted an offer of surgery. Afterwards, although I could move again, the pain remained and could not be explained medically. In my efforts to heal myself, I re-trained and qualified first as a massage therapist then as a sports-massage therapist. I went on to more advanced bodywork techniques and during my studies, I heard about myofascial release and trigger points. Suddenly, everything made sense. I learned how trigger points can be found in all areas of the body and cause a wide range of seemingly unrelated symptoms and ever since I have made it my mission to pass on techniques for self-treating myofascial trigger points.

This book provides information on many common trigger points and the patterns of pain and other symptoms associated with them. We steer clear of complex medical jargon in favour of clear and simple step-by-step guides to self-treat trigger points effectively. Put simply, this is the book I wish I'd had when I started my journey to recovery from chronic pain. I hope you find it a useful guide to living pain-free.

Amanda Oswald

GETTING STARTED

WHAT ARE TRIGGER POINTS?

Trigger points are the knots you find when you instinctively massage or press an area of your body to relieve tension or pain. These points form in connective tissue (fascia) and are typically tender to the touch, but they can also cause pain and other symptoms, felt either where they are or referred to elsewhere in the body.

IDENTIFYING TRIGGER POINTS

Most often, trigger points occur within muscles, but they can also be found in ligaments, around joints, or in the tissues that wrap around nerves and bones. Some can be as large as a golf ball, while others are as tiny as a grain of sand.

Trigger-point therapy involves applying direct and sustained pressure onto trigger points in order to release them. This is distinct from myofascial release, which is a gentle hands-on therapy that works to release restrictions in the fascia, the main connective tissue in the body. However, both use a similar approach to resolve referred pain patterns.

Just as the location and size of trigger points can vary enormously, so can the symptoms caused by them. Also, the connection between a trigger point that needs attention and the symptoms a person experiences is not always obvious.

While muscle pain and loss of movement can affect the area immediately around a particular trigger point, often the pain is felt in a different part of the body – a phenomenon known as "referred pain". The symptoms of a problem trigger point are not always muscular – they can also include problems with vision, sleep, co-ordination, and balance.

Trigger points don't fit neatly into one medical specialism or another, so their role in causing symptoms is often overlooked. Often, symptoms caused by a trigger point can be misdiagnosed as a better-known medical condition.

Finding trigger points
There is no precise map to a trigger point – they occur in different locations in different people. Body maps like this will help you work systematically to locate problem areas.

In this book, the likely areas in which to find trigger points are marked with a dotted line

HOW TRIGGER POINTS OCCUR

Nearly everyone experiences symptoms from trigger points at some time in their lives – whether as a result of trauma, overworked muscles, or the simple wear and tear of daily living. Here are some of the most common causes of trigger points in the body.

SURGERY AND SCAR TISSUE
All surgery creates scar tissue; some of it visible on the skin's surface, some of it internal. Scar tissue can develop into restrictions that cause trigger points.

ACCIDENTS
Any trauma, from a minor scrape to a major accident, can damage muscles and soft tissue. As tissue heals, this can create restrictions which then develop into trigger points.

Tracking down causes
For most people, their trigger points are the result of one or more of the factors on this page.

UNDERUSE OR OVERUSE

Long hours at the computer can cause muscles to become tight and stuck, as can a demanding exercise routine – and tight muscles lead to trigger points.

POSTURE

When our bodies are not in perfect balance, our muscles have to work harder to keep us upright and mobile. This can lead to muscle fatigue, soft-tissue restrictions and, over time, trigger points.

STRESS

Stress is a natural part of life – indeed, some stress can be good for us. However, if our muscles are permanently braced for danger, they can become tight and restricted, causing trigger points to form.

PREPARING TO TREAT TRIGGER POINTS

Trigger points can be caused by many things – usually a combination of work, leisure activities, and other factors such as posture and stress levels. Trigger points form over time and persist if you continue to do the same things in the same way. Now that you know more about the possible causes of trigger points, you can start to do things differently to help yourself.

STARTING SELF CARE

Self care begins the process of undoing the restrictions that have caused your painful trigger points. Knowledge is power – the more you know about trigger points, the more you can do to treat and prevent them. This book is designed to give you a better understanding of what trigger points are and where to find them. All the common trigger points are listed by body area, and for each trigger point there is a diagram of its muscle location – as well as summaries of the most common pain problems as a guide to which trigger points may be causing them. Take time to absorb the information before you embark on your self-care programme.

UNDERSTAND REFERRED SYMPTOMS

For each trigger point there is a predictable pattern of pain and other symptoms. Understanding these patterns will help you give the best self care.

THIS IS NOT AN EXACT SCIENCE

Finding trigger points involves some trial and error – they may not always be in exactly the same spot for everyone.

HEAT BEFORE YOU TREAT

When your body is cold, trigger points are tighter and harder to treat. Warm your body first with exercise or by applying damp heat, for example with a hot-water bottle wrapped in a damp towel.

BE PATIENT

Your trigger points may have developed over many months or years. Be prepared to be patient when working with your trigger points, knowing that regular attention will bring real and lasting benefits.

SELF-CARE PRACTICE

Undertaken regularly, self care can address existing trigger points and prevent new ones from forming. While each exercise in this book has specific instructions, here are some general principles to follow, to ensure you get the most from your self-care work.

01

Get the feel for trigger points
They can feel like grains of sand, small knots, or larger lumps. They will always feel tender to the touch, and some will cause referred pain or twitching.

02

Understand your aims
When you work on a trigger point, you are aiming for a softening and reduction of pain. You may experience immediate relief, but it may take until the following day to feel a benefit.

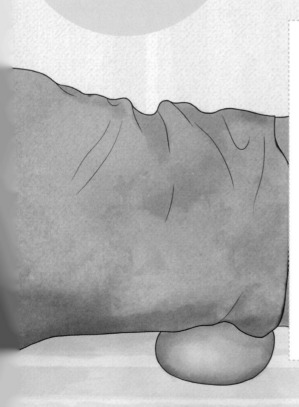

04

Stretch after treatment
Stretching gently after working on a trigger point helps the muscles return to a relaxed state, and prolongs the benefits. Avoid strenuous activity directly after trigger-point work.

03

Little and often is best
If you try to do too much too soon, this can lead to irritation and more pain in the short term. Work on just a few knots at a time and for only up to 30 minutes a day.

HOW TO AVOID MORE TRIGGER POINTS

Trigger points can form as a result of habits that you may not even be aware of. Some things to look at and consider changing are:

- Your work-station setup (at home or in the office)

- Your driving-seat position

- The position in which you sleep

- Your exercise and leisure activities – or lack of them

- How you carry bags or briefcases (for instance, if you always carry a bag on the same shoulder) and how heavy they are.

HOW TO WORK TRIGGER POINTS

Trigger points can be painful and cause surprising symptoms. But self care is easy and can be done safely, following the tips and techniques outlined in this book. Make sure always to take your time and listen to your body as you work.

01

Be creative – it's fine to work on trigger points using your hands, fists, thumbs, fingers or elbows. Use whatever enables you to comfortably maintain the pressure you need to ease the trigger point.

03

Start gently and build up. Gentle pressure encourages your tissues to let go. Going in too hard and fast will only cause tissues to tighten more.

02

Work little and often. Regular daily work brings the best results. Aim for a maximum of 30 minutes per day, either spaced out in sessions or in a single block.

04

Stay within your comfort zone. Using a scale of 0 to 10, where 0 is no pain and 10 is excruciating, never allow the treatment to take you to more than 7.

05

Feel your way. Refer to the trigger point pages to decide which area to work on. Feel around the muscle until you find the point that feels tight and tender to the touch.

08

Sometimes, there won't be an immediate change. If there is no improvement by the next day, try again or move on to a new trigger point. If static pressure is uncomfortable, try deep stroking or kneading – about 10 short strokes per trigger point.

06

When you locate a trigger point, you may notice any of the following: the muscle twitches under your fingers; the area feels tender; referred pain or other familiar symptoms.

09

Finish each session with some gentle stretches of the area you've worked on. Try visualizing the trigger point knot "untangling" itself.

07

Maintain pressure for up to 90 seconds, until either the pain reduces significantly or the trigger point softens or dissolves. After 90 seconds, release the pressure, even if there is no change.

TRIGGER-POINT BALLS AND TOOLS

It's not always easy to reach the trigger point you need to access, or to maintain the required pressure. Many people find tools helpful – either improvised from everyday objects or purpose-made tools you can buy from specialists. Here are some of the tools you might find useful in your self-care work.

USING TOOLS

A specialist trigger-point massaging tool looks a bit like a walking stick with added knobs.

It allows easier access to hard-to-reach areas, such as the shoulder blades. This type of tool is especially useful for those with limited strength in their hands, so that applying pressure is difficult, and for those with mobility problems.

USING BALLS

Balls can be an excellent aid for areas that are harder to reach with hands and fingers. For example, you can place a ball against a wall or the floor and lean into it to work on a trigger point in your back. For working on either side of the backbone, two balls placed in a bag are ideal – you can treat the area without pressing on the sensitive spine itself.

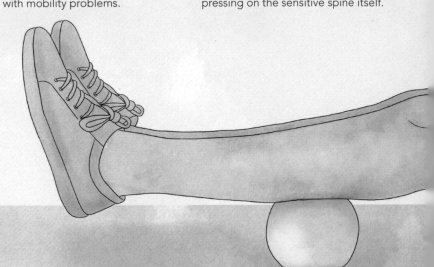

Tennis and squash balls

Tennis balls work well for many people, as they are a good size for individual trigger points. A smaller, harder ball, such as a squash ball, can be useful for more focused pressure. You can also buy balls of different sizes and hardness from specialist retailers.

Inflatable balls

Inflatable myofascial balls are slightly larger than tennis balls and have more "give". They create a gentler, more diffuse pressure. You can buy these balls in a kit which includes a bag (for working with two balls together) and a pump to inflate them.

TRIGGER-POINT TREATMENTS

HEAD
AND NECK

Headaches, migraines, and associated neck pain are among the most common pain problems that people complain of. The trigger points that cause pain in this area can be found in the head and neck, as well as in other parts of the body, such as the shoulders and upper back.

HOW TRIGGER POINTS DEVELOP

• Strain on the neck, particularly if you spend a lot of time at a computer, can be a key cause of trigger points that produce head and neck pain.
• Tightened muscles, often as a result of a head injury, can form trigger points that lead to pain in this area.

SYMPTOMS OF TRIGGER POINTS

• Head pain that can be described as throbbing, sharp, pressure, stabbing pain through the eye, or a dull ache.
• Migraines: symptoms include nausea, vomiting, sensitivity to light and sounds, visual changes, vertigo, or, in the case of silent migraines, no actual pain.
• Stiffness when moving your neck, especially turning it from side to side.

• Tension headaches which feel like a tight band either around the head or from the neck over the top of the head.
• Cluster headaches, usually on one side of the head, which often cause a sharp pain accompanied by other symptoms such as eye watering. These often recur in patterns, which can sometimes be linked to changes in the season.
• NDPH – new daily persistent headache – headaches which suddenly start for no apparent reason, and then continue daily.
• Torticollis, otherwise known as wry neck; a painful crick in the neck typically from sleeping in an awkward position.
• Pain that develops following a whiplash injury, sometimes weeks or months after the original incident.

Pain patterns

The location and nature of your head and neck pain can indicate the most likely places to find the trigger points responsible.

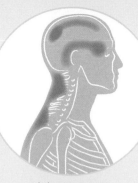

Likely trigger points:
BACK OF NECK, pages 28–29

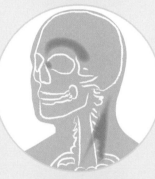

Likely trigger points:
SIDE OF NECK 1, pages 30–31

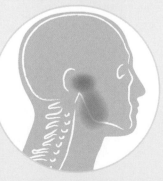

Likely trigger points:
TEMPLES, pages 44–45

Continued ▶▶

Trigger-point locations

Use these body maps to
locate likely trigger points
and navigate to
pages that describe
how to treat them.

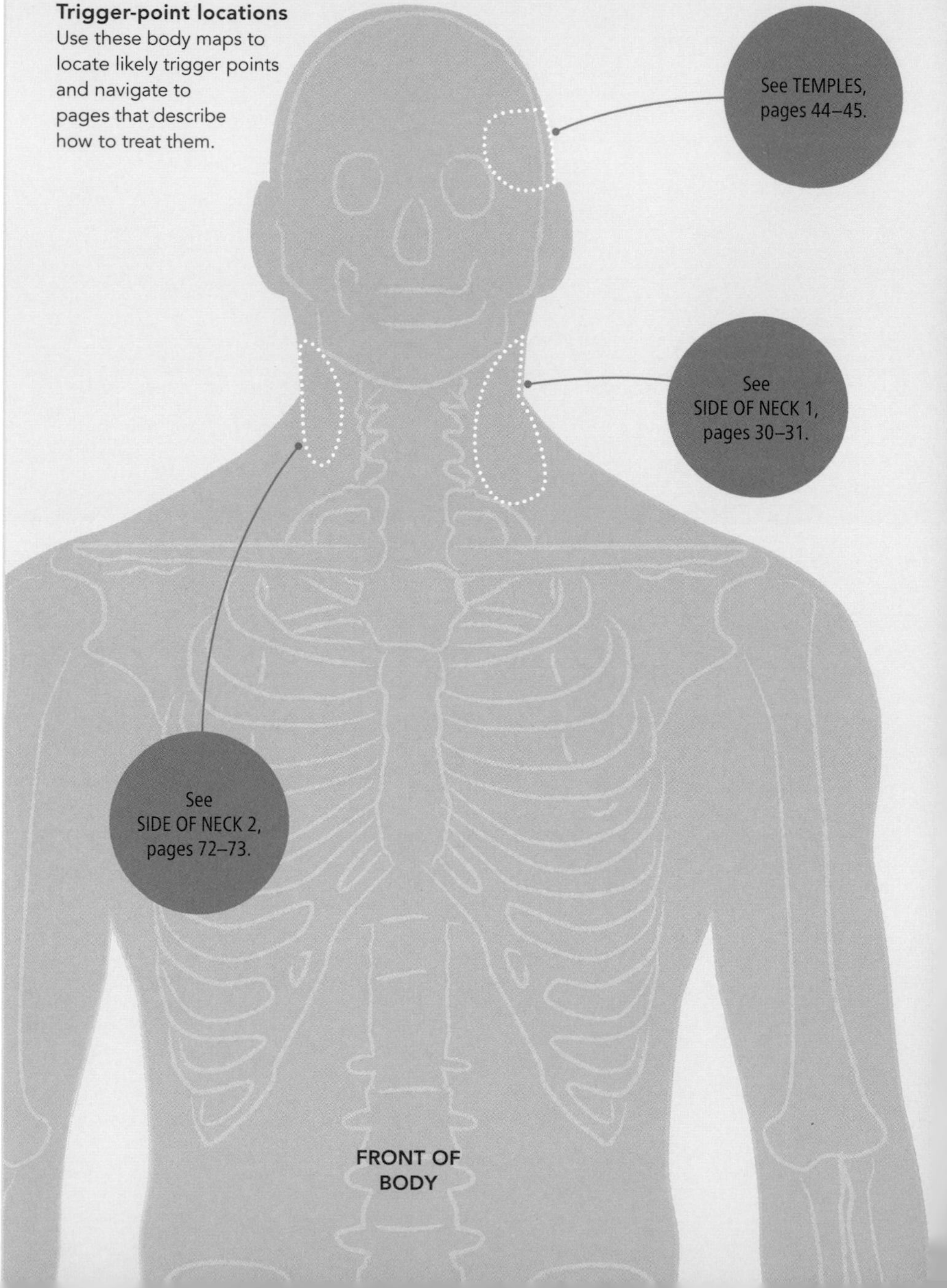

See TEMPLES,
pages 44–45.

See
SIDE OF NECK 1,
pages 30–31.

See
SIDE OF NECK 2,
pages 72–73.

**FRONT OF
BODY**

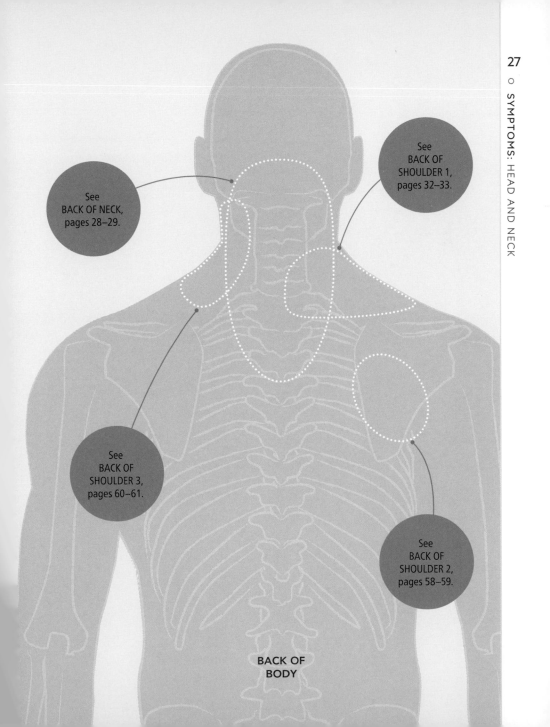

See
BACK OF
SHOULDER 1,
pages 32–33.

See
BACK OF NECK,
pages 28–29.

See
BACK OF
SHOULDER 3,
pages 60–61.

See
BACK OF
SHOULDER 2,
pages 58–59.

**BACK OF
BODY**

BACK
OF NECK

Several muscles from the spine and the neck attach to the base of the skull at the back of the neck. Trigger points in this area can cause a stiff, painful neck, headaches, and burning pain in the scalp. Also, trigger points in eye-movement muscles under the base of the skull may result in eye pain and disruption of vision.

03

Start by placing the balls just below the base of your skull. Position one on either side of your upper spine for comfort.

02

Lie flat on the floor or on a bed with a cushion under your head.

01

Use two inflatable balls in a bag (see page 20) to give the gentle level of pressure needed for this sensitive area.

POSSIBLE PAIN PATTERNS

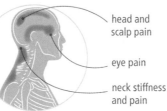

head and scalp pain

eye pain

neck stiffness and pain

MUSCLE LOCATOR

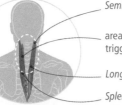

Semispinalis muscle

area of possible trigger points

Longissimus muscle

Splenius muscle

04

Rest against the balls. The weight of your neck and head will be sufficient to release trigger points. Stop when you feel a sense of release, or after 90 seconds.

05

To release lower trigger points, move the balls further down a little, and repeat step 4.

Rest gently on the balls without pushing down

SIDE OF NECK 1

The *Sternocleidomastoid* muscle attaches to your breastbone, collarbone, and mastoid process behind the ear. There is one muscle on either side of your neck and each turns your head to the opposite side, for example, when you look over your shoulder. Trigger points here cause tension headaches, pain in the sinuses and throat, and contribute to dizziness, tinnitus, earache, jaw pain, blocked sinuses, and a dry cough.

POSSIBLE PAIN PATTERNS

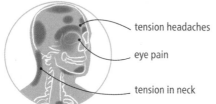

tension headaches

eye pain

tension in neck

MUSCLE LOCATOR

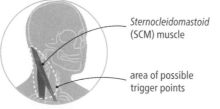

Sternocleidomastoid (SCM) muscle

area of possible trigger points

03

Take the SCM into a firm pincer grip. It can feel like a slippery rope and difficult to grasp, but with practice you will get the feel of it.

02

Grasp the muscle itself between the fingers and thumb of the hand opposite the side you are working.

01

Relax the SCM muscle by turning your head to the side you want to work on first. Tilt your head slightly towards your shoulder.

04

Work along the muscle from the base upwards. Whenever you feel a tender point, hold with a gentle pinch until the tenderness eases, or for about 90 seconds.

05

To work the other SCM, turn your head in the opposite direction and repeat steps 2–4. Repeat daily for as long as you need to.

Grasp only the muscle, not the whole neck

BACK OF SHOULDER 1

Trigger points found in the upper part of the *Trapezius* muscle can cause tension headaches and headaches stretching around the ear in the shape of a question mark. They can also contribute to jaw and neck pain by creating tension in your neck. You can work these trigger points with your hand as shown here, or with a ball placed against a wall (see page 20).

02

Pinch the muscle between your fingers and thumb to find trigger points. A trigger point will feel tender and pressing it may recreate a headache, jaw pain, or neck tension.

01

Using the hand opposite your painful side, feel for the upper *Trapezius* muscle, which sits over the top of your shoulder. Place your fingers on your shoulder near the base of your neck.

POSSIBLE PAIN PATTERNS

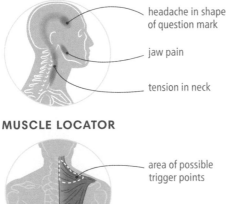

headache in shape of question mark

jaw pain

tension in neck

MUSCLE LOCATOR

area of possible trigger points

Trapezius muscle

03
___

When you find
a trigger point, hold the
compression here for up to
90 seconds; if pressing causes
pain, you can stop if this
recedes before the full
90 seconds.

04
___

Work along the muscle
to locate further trigger
points and, if you find
them, treat them by
repeating step 3.

Trigger points may feel
like small, hard knots

JAW, EAR, AND FACE

The most common cause of facial pain is a group of conditions known as temporo-mandibular (jaw) joint and muscle disorders (TMJDs), which is why face and jaw problems are commonly referred to as TMJ. Pain in these areas is most often caused by trigger points in the muscles of your face, jaw, mouth, and neck.

HOW TRIGGER POINTS DEVELOP

• For their size, the jaw muscles are the strongest in the body. Clenching your jaw, often as a result of stress, can cause trigger points to form in this area.

• Bruxism – clenching or grinding your teeth at night – can cause trigger points to develop in the jaw. Wearing a splint to protect your teeth can change your bite and cause further trigger points.

• Holding your mouth open for long periods of time, such as during prolonged dental work, can cause trigger points.

• Trigger points in your jaw can refer pain and other symptoms in a wide pattern that extends into your head and ears.

SYMPTOMS OF TRIGGER POINTS

• TMJ and jaw pain, including very intense pain felt in one or both of your jaw joints, especially when chewing food.

• Clicking and popping noises in your jaw and difficulty opening and closing your mouth.

• Facial pain and other symptoms such as blurred vision, eye watering, blocked sinuses, and numbness.

• Tinnitus, which is noise in your ears variously described as ringing, buzzing, humming, grinding, whistling, or whooshing. Tinnitus is often worse at night and, for some, when they hold their head in a particular position. For others, the condition is tension-related.

Pain patterns

Evaluate the type and location of pain in the jaw, ear, and face to help you locate any trigger points that could be the cause.

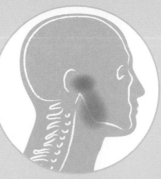

Likely trigger points:
JAW, pages 38–39

Likely trigger points:
INSIDE MOUTH 1, pages 40–41

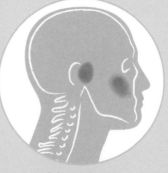

Likely trigger points:
INSIDE MOUTH 2, pages 42–43

Likely trigger points:
TEMPLES, pages 44–45

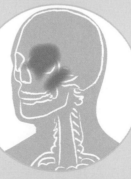

Likely trigger points:
FACE, pages 46–47

Continued ▶▶

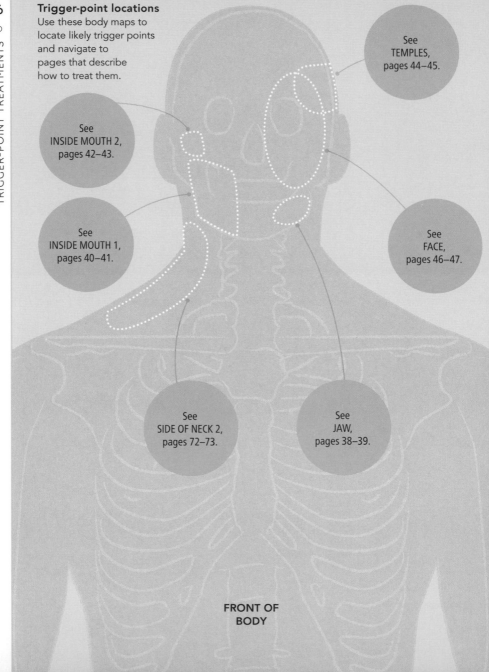

Trigger-point locations
Use these body maps to
locate likely trigger points
and navigate to
pages that describe
how to treat them.

See
TEMPLES,
pages 44–45.

See
INSIDE MOUTH 2,
pages 42–43.

See
INSIDE MOUTH 1,
pages 40–41.

See
FACE,
pages 46–47.

See
SIDE OF NECK 2,
pages 72–73.

See
JAW,
pages 38–39.

**FRONT OF
BODY**

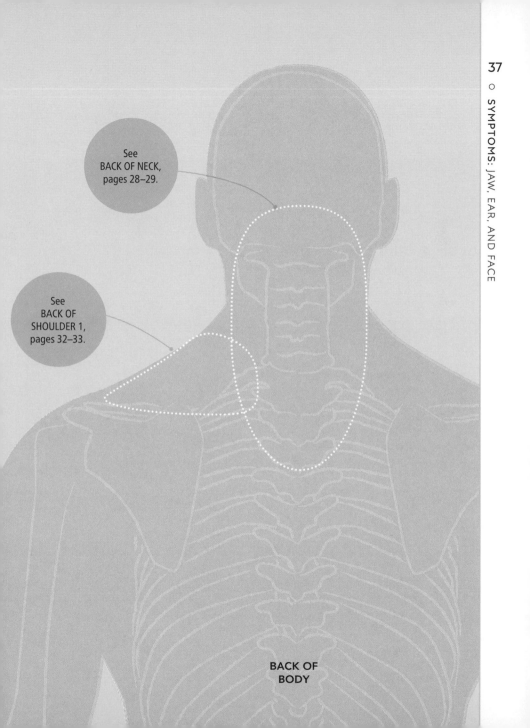

See
BACK OF NECK,
pages 28–29.

See
BACK OF
SHOULDER 1,
pages 32–33.

BACK OF
BODY

JAW

The muscles around the jaw can get very tight, especially from stress or extensive dental work. The two *Medial pterygoid* muscles sit just inside the left and right lower jawbones, mirroring the position of the *Masseter* (see pages 40–41) on the outside. Trigger points here contribute to jaw, TMJ, throat, and mouth pain. They can also make it difficult for you to open your mouth fully.

POSSIBLE PAIN PATTERNS

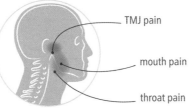

TMJ pain

mouth pain

throat pain

MUSCLE LOCATOR

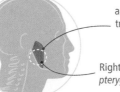

area of possible trigger points

Right *Medial pterygoid* muscle

03

Press your thumb directly up into the soft tissue on the inside of the bone. Increase pressure gently and gradually, as this area can feel extremely tender.

02

Come back towards your chin slightly until you feel a notch in the bottom of the bone.

01

On the same side as your pain, use your thumb to feel along your jawbone for the corner, just below your ear.

04

If pressing on this area makes your mouth water, you've gone too far along the bone towards the chin and are on a salivary gland, so move back.

05

Continue the pressure until you feel a sense of softening. Stop after 90 seconds if this change doesn't occur.

Tilt your head slightly towards your shoulder

39

○ SYMPTOMS: JAW, EAR, AND FACE

INSIDE MOUTH 1

The *Masseter* muscle enables you to chew, and for its size it is the strongest muscle in the body. Trigger points here can cause symptoms including TMJ pain, toothache, jaw, face, or ear pain, headaches, and itchy ears. Trigger points here can be brought on by bruxism (tooth grinding and clenching at night). Conversely, this condition can be a symptom of trigger points here and in other face muscles.

POSSIBLE PAIN PATTERNS

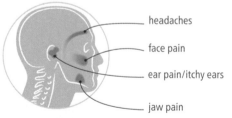

— headaches

— face pain

— ear pain/itchy ears

— jaw pain

MUSCLE LOCATOR

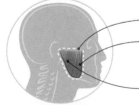

Masseter muscle

area of possible trigger points

temporomandibular joint (TMJ)

03

To locate the Masseter muscle, clench your jaw briefly and you'll feel the muscle bulging just in front of the bone.

02

Keeping your jaw loose, slide your thumb into your mouth between your teeth and your cheek. Take your thumb back as far as it will go to the jawbone.

01

Use the hand opposite the painful side that you will be working on.

04

Loosen your jaw, then gently and slowly massage the whole muscle between your thumb and fingers.

05

As you move around the muscle, stop at any tender points and press gently either until the tenderness eases, or for 90 seconds.

Finger pads rest on the outside of the cheeks

Thumb pad inside mouth faces cheek

INSIDE MOUTH 2

The name *Lateral pterygoid* derives from the Greek for wing, and this wing-shaped muscle is tucked away under the top corner of your cheekbone. Trigger points here can contribute to TMJ and jaw pain (see page 34) and make it difficult for you to open your mouth fully. They can also cause face pain, and clicking and popping noises in your jaw.

POSSIBLE PAIN PATTERNS

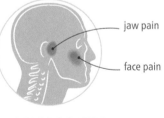

— jaw pain

— face pain

MUSCLE LOCATOR

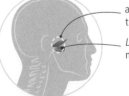

— area of possible trigger points

— *Lateral pterygoid* muscle

03

To locate the muscle, slide your finger up to the back corner of your cheekbone where you will find a small pocket of soft tissue tucked under the bone.

02

Keeping your jaw loose, slide your index finger into your mouth between your teeth and your cheek.

01

Make sure your nails are short before you start. Use the hand on the same side as the painful area you are working on.

04

Press up into this pocket. Be gentle – if you have trigger points here, the area can feel quite tender. When you locate a trigger point, keep pressing until it softens, or for about 90 seconds.

Press gently upwards with tip of index finger

05

Continue gently probing the muscle, and when you locate more trigger points, repeat step 4.

TEMPLES

Your temples are home to large, fan-like *Temporalis* muscles that extend down on either side of your face, to your jaw. These chewing muscles can become very tight, sometimes to the point that they feel ridged when massaged. Trigger points here can form due to stress, the chewing of gum, or prolonged dental work. Symptoms include head, tooth, and jaw pain, sensitive teeth, and eyebrow pain.

02

Use your knuckles to probe the area gently for trigger points. Move your knuckles slowly, in circular or brushing strokes.

01

Sit with your elbow supported on a chair arm or a table. Form a soft fist with your hand and rest your temple area on your knuckles.

POSSIBLE PAIN PATTERNS

headaches on side of head

toothache

jaw pain

MUSCLE & TRIGGER POINT LOCATOR

Temporalis muscle

area of possible trigger points

03

When you find a tender spot, allow the weight of your head to sink deeper onto your knuckles. Remain in this position until the tenderness eases, or for 90 seconds.

make a soft fist

Move along temple from front to back

04

Carry on working your knuckles over the temple area to locate and treat other possible trigger points.

FACE

The area around your eyes and nose can be prone to trigger points, which can cause searing headaches just above the eye, sinus pain, and pain across the bridge of the nose. Other trigger-point symptoms can include eye strain, drooping eyelid, and disturbance to your vision. Trigger points here can also cause allergy-type symptoms such as a runny nose, sneezing, and itchy eyes.

02

When you find a tender point, apply a sustained, gentle pressure on the trigger point.

01

Work one side of the face at a time. Starting under one eye, use your fingers to press gently into the sensitive areas all around your eye socket and down the side of your nose.

POSSIBLE PAIN PATTERNS

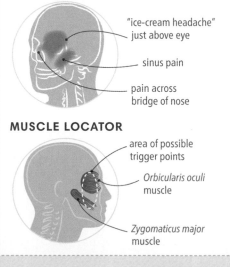

"ice-cream headache" just above eye

sinus pain

pain across bridge of nose

MUSCLE LOCATOR

area of possible trigger points

Orbicularis oculi muscle

Zygomaticus major muscle

03

Maintain pressure until you feel the tenderness easing, or for around 90 seconds.

04

Continue moving round the eye. Be aware that pressing the top corner of your eye socket may set off a temporary headache – this clearly indicates you are directly on a trigger point.

Extreme tenderness indicates trigger point

05

Repeat steps 1–4 for the other eye. It's important to work both sides, since trigger points on one side of the face can refer pain to the other side.

CHEST

Trigger points in the muscles and soft tissues of your chest, side, and neck can be the cause of considerable pain and discomfort in the chest. Unexpected chest pain can be alarming. If you do experience pain in this area, you should first consult a medical professional to rule out any underlying causes.

HOW TRIGGER POINTS DEVELOP

• Any condition that makes it difficult to breathe will increase the strain on muscles in your back, rib area, and neck – all of which can develop trigger points.

• Costochondritis (inflammation of cartilage on the side of the breastbone) can cause trigger points in the area.

• Carrying a bag on one shoulder can lead to tension and trigger points in the muscles on the side of your torso.

• Repetitive movements, such as during sport, can strain the chest muscles and cause trigger points to form.

• Over-exertion such as heavy lifting or compression from a seat belt in a car accident can develop into trigger points.

SYMPTOMS OF TRIGGER POINTS

• Sharp shooting pains.

• Difficulty taking deep breaths.

• General pain and sensitivity that can be felt deep inside the chest.

• A sense of constriction, especially during exertion.

• A chronic cough which persists after the infection that caused it has healed.

PAIN PATTERNS

The location and nature of your pain can indicate likely places to look for the trigger points responsible.

Likely trigger points:
CHEST 1, pages 50–51

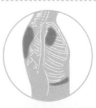

Likely trigger points:
SIDE OF CHEST, pages 52–53

Trigger-point locations

Use the body map to locate
trigger points and navigate
to the pages that describe
how to work them.

See
SIDE OF NECK 1,
pages 30–31.

See
CHEST 2,
pages 76–77.

See
CHEST 1,
pages 50–51.

See
SIDE OF CHEST,
pages 52–53.

See BETWEEN
THE RIBS,
pages 100–101.

FRONT OF BODY

BACK OF BODY

CHEST 1

The chest contains two principal muscles. The *Pectoralis major*, or main "pecs" muscle, gives your chest shape, and the smaller muscle underneath is known as the *Pectoralis minor*. This muscle is significant because it can trap your arm nerves against your ribs when it is tight. Trigger points in this area can cause pain in your chest and the whole arm, as well as pain and numbness in your forearm, hand, and fingers.

PAIN PATTERN

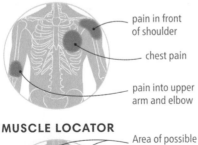

pain in front of shoulder

chest pain

pain into upper arm and elbow

MUSCLE LOCATOR

Area of possible trigger points

Pectoralis minor muscle

Pectoralis major muscle

03

Place the ball against the painful side of your chest and either lean into the wall or lie with the ball between your body and the floor or bed.

02

Stand at the corner of two walls, with your painful side nearest the wall. Alternatively, lie face down on the floor or on a bed.

01

Use an inflatable ball (see page 20) to give the gentle level of pressure needed for this sensitive area.

04

Move the ball slowly between your shoulder and breastbone to look for tender points. When you find a trigger point (it will feel tender), lean into the ball to apply gentle pressure.

Lean into ball when trigger point located

05

Hold the position for about 90 seconds, or until you feel a sense of ease. Repeat step 4 to locate and work more trigger points.

SIDE OF CHEST

Trigger points in this area, which forms part of your armpit, cause a range of symptoms from a stitch when running to soreness in the area and painful breathing. Trigger points can develop as a result of conditions such as a chronic cough, asthma, or respiratory infections. It is important to seek medical advice since what appear to be trigger points in this area can also be possible symptoms of lung or heart problems.

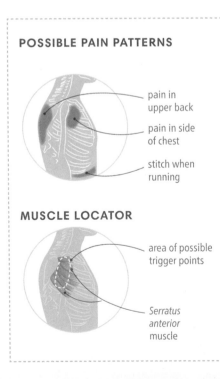

POSSIBLE PAIN PATTERNS

— pain in upper back

— pain in side of chest

— stitch when running

MUSCLE LOCATOR

— area of possible trigger points

— *Serratus anterior muscle*

01

Reach over towards your sore side with the opposite hand. Alternatively, lean on a ball with your side against a wall.

02

Rest your hand on the side of your ribs below your armpit. Starting at the top rib, find tender points by pulling your fingers across the ribs. Use firm, not hard, pressure and move slowly as this area can be very tender.

03

Once you locate a trigger point, apply pressure with your finger pads until you feel it change and the pain eases. Stop after about 90 seconds if it does not.

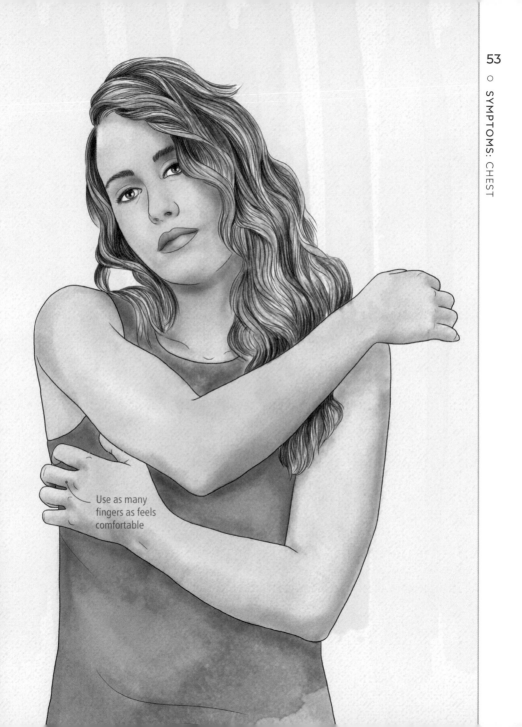

Use as many fingers as feels comfortable

SHOULDERS

The shoulder joint is the most mobile in the body, and shoulders are particularly prone to damage that can lead to trigger points. Trigger points in the muscles and soft tissues of the neck, shoulders, armpits, arms, chest, and back can all contribute to shoulder problems, even though the pain is most commonly felt in the shoulder itself.

HOW TRIGGER POINTS DEVELOP

- The mobility of the shoulder muscles means that they can get damaged more easily by joint dislocation, muscle tears, and sprains, all resulting in trigger points.
- Damage or overuse can cause trigger points to develop in the muscles supporting the shoulder joint.
- Repetitive actions, such as computer work or some sports, can cause strain and pain in the shoulders.
- The rotator cuff muscles control the rotational range of movement of your shoulder. Trigger points can develop when these muscles are damaged by overuse or injury.
- Frozen shoulder is a very common problem in this area. True frozen shoulder occurs when the tissues inside the shoulder joint get stuck together. However, many people who are diagnosed with this actually have pain and restricted movement caused by trigger points.

SYMPTOMS OF TRIGGER POINTS

- Dull aches and pains in the shoulder joint, which can make it impossible to sleep on the affected side.
- Sharp pains which restrict certain movements such as lifting your arm out to the side, putting your arm behind your back, brushing your hair, dressing and undressing.
- A feeling of weakness in your arm when lifting things, and an inability to move your arm in some directions.

Pain patterns
The site and type of shoulder pain can help you identify the best places to look for trigger points causing the pain.

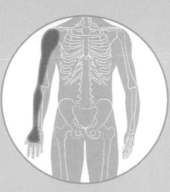

Likely trigger points:
BACK OF SHOULDER 2, pages 58–59

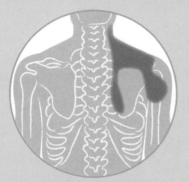

Likely trigger points:
BACK OF SHOULDER 3, pages 60–61

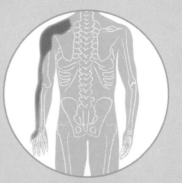

Likely trigger points:
BACK OF SHOULDER 4, pages 62–63

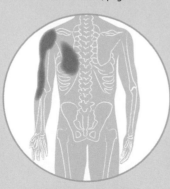

Likely trigger points:
SIDE OF ARMPIT 1, pages 64–65

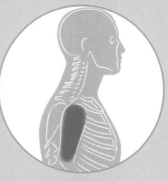

Likely trigger points:
TOP OF SHOULDER, pages 66–67

Continued ▶▶

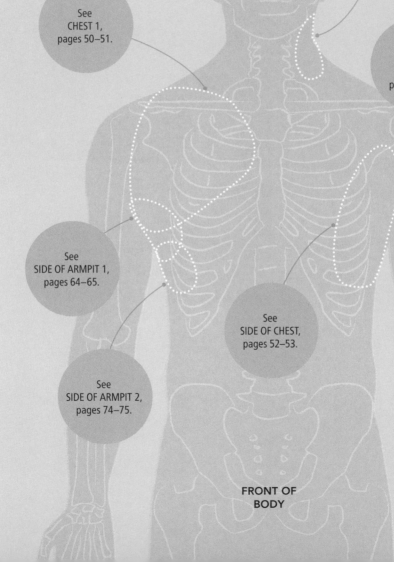

Trigger-point locations
Use these body maps to
locate likely trigger points
and navigate to
pages that describe
how to treat them.

See
SIDE OF NECK 1,
pages 30–31.

See
CHEST 1,
pages 50–51.

See
TOP OF
SHOULDER,
pages 66–67.

See
SIDE OF ARMPIT 1,
pages 64–65.

See
SIDE OF CHEST,
pages 52–53.

See
SIDE OF ARMPIT 2,
pages 74–75.

FRONT OF
BODY

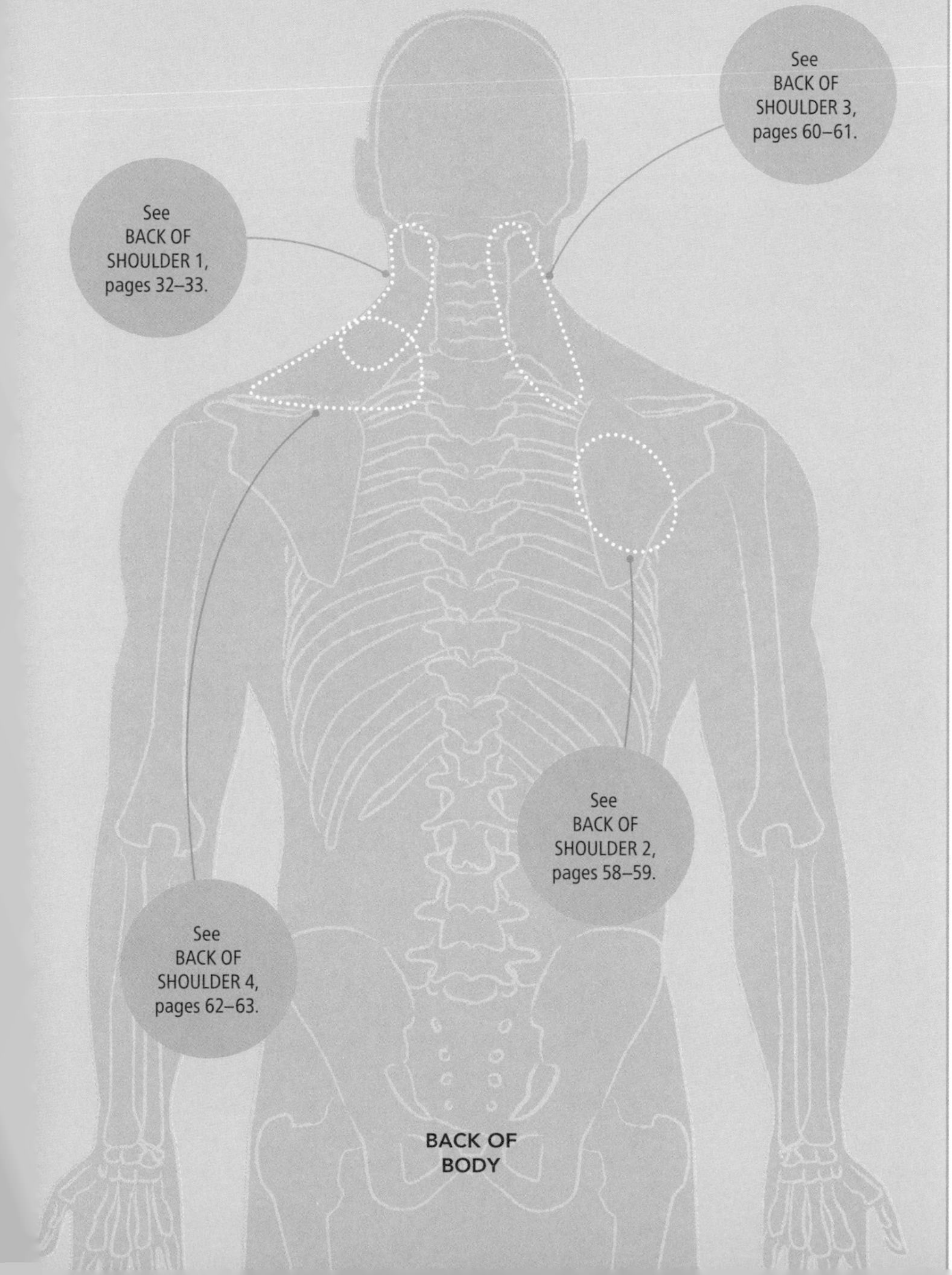

See
BACK OF
SHOULDER 3,
pages 60–61.

See
BACK OF
SHOULDER 1,
pages 32–33.

See
BACK OF
SHOULDER 2,
pages 58–59.

See
BACK OF
SHOULDER 4,
pages 62–63.

**BACK OF
BODY**

BACK OF SHOULDER 2

Trigger points in the flat muscle covering the shoulder blade, which helps to rotate your arm for actions such as brushing your hair, make it difficult to move your arm, particularly to reach behind you. They can also make your arm feel weak and make it painful to sleep on your side. Trigger points here may cause a deep aching pain that spreads down the arm.

POSSIBLE PAIN PATTERNS

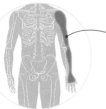

pain in front and side of arm, from shoulder to hand

MUSCLE LOCATOR

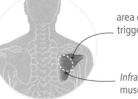

area of possible trigger points

Infraspinatus muscle

03

Once you find a trigger point, lean into it, sustaining gentle pressure for at least 90 seconds, or until you feel a sense of eased tension or reduced pain.

02

Move the ball slowly around the shoulder-blade area. Some, but not all, trigger points when pressed will re-create referred pain down your arm.

01

Lean on a ball against a wall. Try different-sized balls to find one that is comfortable to use in this painful area. Alternatively, you could also use a trigger-point tool (see page 20).

04

Move the ball slowly and systematically across the whole area of the shoulder blade. Repeat step 3 to treat any other trigger points you find.

Referred pain may be felt as you work the shoulder-blade area

BACK OF SHOULDER 3

The *Levator scapulae* is one of the muscles at the back of your shoulder that helps to move your shoulder blade – you use it when you shrug your shoulders. Trigger points here cause pain around the top of the shoulder blade and make it difficult to turn your neck, for example when looking over your shoulder.

02

Move your thumb around the area until you find a tender point. Then, either rub slowly over the spot, or apply pressure until the tenderness eases or after 90 seconds.

01

Work on the side on which you feel pain. To work the top of the muscle, press your thumb into the back of the side of your neck, just below where your hairline starts.

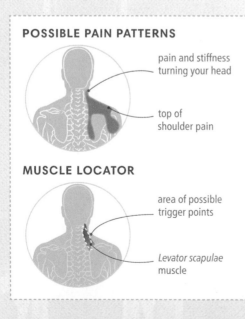

POSSIBLE PAIN PATTERNS

pain and stiffness turning your head

top of shoulder pain

MUSCLE LOCATOR

area of possible trigger points

Levator scapulae muscle

03

To work on the lower part of the muscle, use a ball. A trigger-point ball is easier for pinpointing trigger points, but if the pressure is too hard, use an inflatable ball instead.

04

Place the ball against a wall and lean your lower neck against it. Check the muscle-locator diagram on the left if you're not sure where to position the ball.

05

Move the ball slowly over the area. If you find tender points, lean into the ball and hold the position for 90 seconds or until the pain eases.

Trigger points may be found up behind ear

Keep the thumb flat so the nail doesn't dig into the skin

BACK OF SHOULDER 4

The *Supraspinatus* muscle, which helps to lift your arm, is tucked into a bony pocket at the top of your shoulder blade. Trigger points can make it painful to lift your arm and cause a deep ache from the top of your shoulder down your arm. They can also tighten your arm in the shoulder socket, causing clicking and popping when you move your arm.

POSSIBLE PAIN PATTERNS

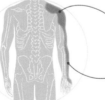

deep ache on outer side of shoulder

pain along length of arm

MUSCLE LOCATOR

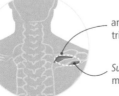

area of possible trigger points

Supraspinatus muscle

02
Confirm you are in the right place by lifting your arm slightly – this will cause the muscle to tighten and bulge beneath your fingers.

01
To locate the muscle, take the hand opposite your sore arm and rest the palm on your collarbone, with your fingers over your shoulder.

03

Using your fingers to work this area can be tiring for them, so you could use a trigger-point tool instead. Place the tool over your shoulder to rest on the work area.

04

If using your fingers, rub the area with slow horizontal or circular strokes to locate trigger points. When you find a tender point, apply gentle pressure to it.

05

Apply the pressure until discomfort eases, or for 90 seconds. Repeat step 4 to find and treat more trigger points.

Apply static pressure with the tool, or massaging strokes if using fingers

SIDE OF ARMPIT 1

Strong muscles in this area move and rotate your arm, especially during sporting activities such as climbing, tennis, and swimming front crawl. Trigger points here can cause mid-back and shoulder pain, making it difficult to reach up above your head and causing a sharp pain in the back of your shoulder when you reach forwards or try to pick something up.

POSSIBLE PAIN PATTERNS

sharp pain in shoulder joint

pain from top of arm to forearm

mid-back pain

MUSCLE LOCATOR

Areas of possible trigger points

Teres minor muscle

Latissimus dorsi muscle

03

Position the ball on the location you found with your hand. Lean your body and ball against the wall.

02

Using your opposite hand, feel along the side of your painful armpit with your fingers to find the side of your shoulder blade, which feels like a bony ridge.

01

You can work this area using a ball, by either lying down or leaning against a wall. Choose the type of ball that feels most comfortable (see page 20).

04

Sink your body into the ball.
If it feels tender, wait for
90 seconds, or until you feel
a sense of ease.

Don't worry about
exactly where to place
the ball: working
anywhere here will
access trigger points

05

Carry on slowly moving
the ball over the area.
Stop at any tender
points you find and
repeat step 4.

TOP OF SHOULDER

The *Deltoid* muscle wraps round the top of your shoulder like a cap. It moves your arm back, sideways, and to the front when, for example, you lift your arm to wave. You also use it when you lift and carry things. Trigger points here are felt mainly when you use the muscle and they can cause a sense of weakness in your arm, and pain in your shoulder.

01

Find trigger points in this area using an inflatable ball. To work on a specific trigger point, try using a trigger-point ball.

02

Position the ball between the top side of your sore shoulder and a wall. Move the ball around slowly until you find a tender area.

03

Once you find a trigger point, lean into the ball to apply gentle pressure. Continue for 90 seconds, or until the tenderness eases.

POSSIBLE PAIN PATTERNS

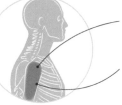

pain felt mainly when using the muscle

upper arm, both back and front, can be affected

MUSCLE LOCATOR

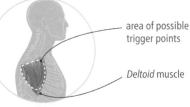

area of possible trigger points

Deltoid muscle

You may find trigger points around the whole of the top of the shoulder

05

For each trigger point you locate, repeat step 3 to ease the tension and tenderness.

04

Turn your body and lean against the ball to find other trigger points on the front, back, and side of your shoulder.

ARMS AND HANDS

Arm, elbow, and hand problems are among the most common work-related health issues. Prolonged, repetitive use of a keyboard causes the muscles and soft tissues to become tired and irritated, leading to a number of conditions commonly grouped together as Repetitive Strain Injury (RSI). Trigger points are the main cause of RSI.

HOW TRIGGER POINTS DEVELOP

• Overuse of the hands and fingers, such as when using computers and mobile devices, can cause trigger points to form.
• Sport and leisure activities that involve repetitive hand or arm movements, such as golf, tennis, sewing, or playing a musical instrument, can all lead to trigger points.
• Trigger points in this area are often misdiagnosed as carpal tunnel syndrome.

SYMPTOMS OF TRIGGER POINTS

• Sharp, shooting pains or dull aching.
• Tiredness and swelling, especially in your forearms.
• Tingling, or pins and needles, mainly in forearms and fingers.
• Numbness and loss of grip strength, affecting actions such as writing or turning a door knob.
• Burning elbow pain, making it painful to lift or grip things. Often referred to as golfers' or tennis elbow.

Pain patterns
The location and nature of your arm and hand pain can indicate the most likely place to find the trigger points responsible.

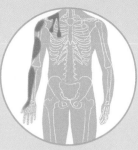

Likely trigger points:
SIDE OF NECK 2, pages 72–73

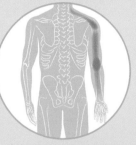

Likely trigger points: SIDE OF
ARMPIT 2, pages 74–75

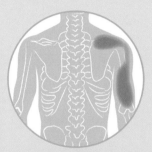

Likely trigger points: CHEST 2,
pages 76–77

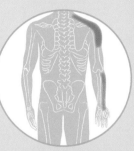

Likely trigger points: FRONT OF
UPPER ARM, pages 78–79

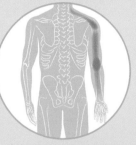

Likely trigger points: BACK OF
UPPER ARM, pages 80–81

Likely trigger points: BACK OF
FOREARM, pages 82–83

Likely trigger points: FRONT OF
FOREARM, pages 84–85

Likely trigger points: PALM OF
HAND, pages 86–87

Continued ▶▶

Trigger-point locations
These body maps will help
you to locate likely trigger
points, then navigate to
the pages that describe
how to treat them.

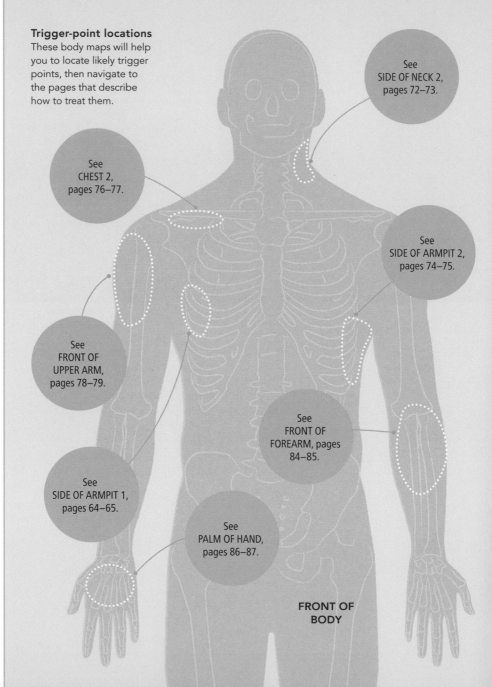

See
SIDE OF NECK 2,
pages 72–73.

See
CHEST 2,
pages 76–77.

See
SIDE OF ARMPIT 2,
pages 74–75.

See
FRONT OF
UPPER ARM,
pages 78–79.

See
FRONT OF
FOREARM, pages
84–85.

See
SIDE OF ARMPIT 1,
pages 64–65.

See
PALM OF HAND,
pages 86–87.

**FRONT OF
BODY**

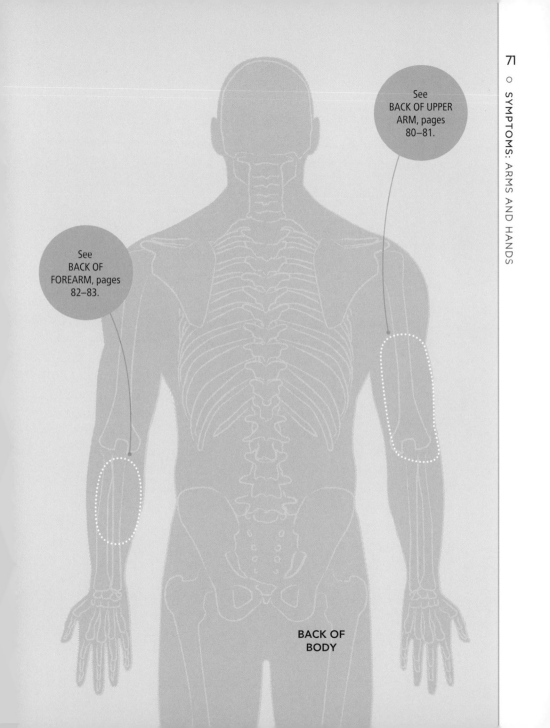

See
BACK OF UPPER
ARM, pages
80–81.

See
BACK OF
FOREARM, pages
82–83.

**BACK OF
BODY**

SIDE OF NECK 2

Trigger points in the side of your neck can affect your arms because this is where the arm nerves exit from the spinal cord on the start of their journey to your arms. Trigger points here can cause symptoms of repetitive strain injury (RSI) and loss of strength in your arms and hands, and are typically caused by prolonged, repeated movements of your arms and hands, such as typing at a computer keyboard.

POSSIBLE PAIN PATTERNS

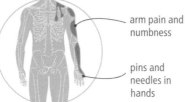

arm pain and numbness

pins and needles in hands

MUSCLE LOCATOR

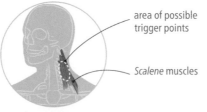

area of possible trigger points

Scalene muscles

02

Place a ball under your neck towards the middle. An inflatable ball is best, as it exerts very gentle pressure in this sensitive area.

01

Lie on the floor or on a bed, on the same side as your symptoms. Place a small cushion under your head for comfort.

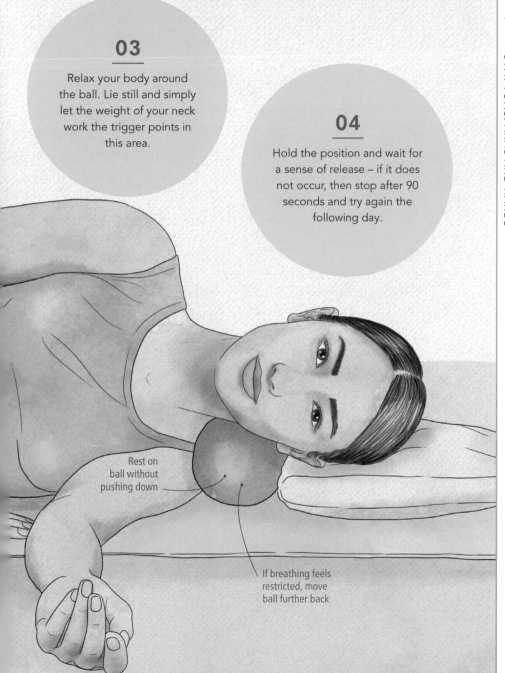

03

Relax your body around the ball. Lie still and simply let the weight of your neck work the trigger points in this area.

04

Hold the position and wait for a sense of release – if it does not occur, then stop after 90 seconds and try again the following day.

Rest on ball without pushing down

If breathing feels restricted, move ball further back

SIDE OF ARMPIT 2

The muscles in the area around your armpit can become tight with overuse - working at the computer with your arms held in front of you for prolonged periods is a common cause of issues here. Trigger points can cause restrictions in your arm movement and pain, both in the back of your shoulder and down the back of your arm.

POSSIBLE PAIN PATTERNS

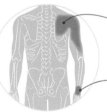

pain deep in the back of shoulder

pain down back of arm to wrist

MUSCLE LOCATOR

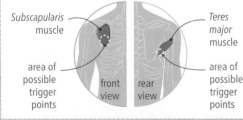

Subscapularis muscle

area of possible trigger points

front view

rear view

Teres major muscle

area of possible trigger points

03

Place an inflatable ball at the back of your armpit, on the tender trigger point that you located with your hand. Lie on the ball and relax your body around it.

02

Feel along the side of your armpit with your other hand to find the side of your shoulder blade, which feels like a bony ridge. Feel with your fingers for any points that are tender.

01

Lie on the floor or a bed, on your painful side. Raise your arm on that side above your head. If you find this difficult, lean on a ball against a wall instead (see page 20).

04

Rest on the tender point, without pushing down, until it eases, or for about 90 seconds. Treat other trigger points in the same way.

05

To work the top of your armpit, move the ball to rest on other trigger points you locate with your hand, and repeat step 4 to ease them.

Support your head with a pillow on your arm

CHEST 2

The name of the *Subclavius* muscle indicates its location, which is directly under (*sub*) your collarbone (*clavius*). For its size, this tiny muscle can cause a disproportionate number of problems. Trigger points here cause pain in your arm and into your hand and fingers. When the muscle is tight, for instance as a result of a lot of computer work, it contributes to pulling your shoulder forwards.

03

Working very carefully, push your finger gently back up towards the bone. You will know when you find a trigger point here as it will feel very tender.

02

Sink your middle finger in under your collarbone. If the area feels very tight, create space by rounding your shoulder to bring it forwards.

POSSIBLE PAIN PATTERNS

pain just below collarbone

pain in your arm and down into fingers

MUSCLE LOCATOR

Subclavius muscle

area of possible trigger points

01

Take the hand opposite the painful side across your body, resting your fingers on the collarbone.

04

Maintain gentle pressure until you feel a sense of release, or stop after 90 seconds if this doesn't happen.

Feel for very tender area under middle of collarbone

FRONT OF UPPER ARM

You may have trigger points in the front of your upper arm caused by repetitive everyday arm movements, although you will rarely experience pain where these trigger points occur. Instead, pain will refer to your shoulders, the backs of the arms, forearms, and hands. Trigger points here also sometimes make it difficult to straighten your arm.

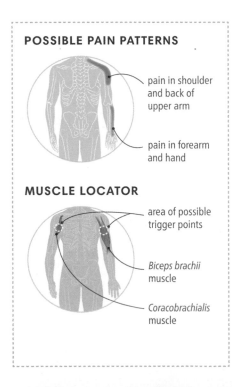

POSSIBLE PAIN PATTERNS

pain in shoulder and back of upper arm

pain in forearm and hand

MUSCLE LOCATOR

area of possible trigger points

Biceps brachii muscle

Coracobrachialis muscle

01

Form a soft fist with the hand opposite your painful side. Place your knuckles near the top of your sore arm.

02

Slowly and gently "rake" your fist down the front of the arm towards the elbow. Pause when you encounter a tender area. Apply pressure until it eases, stopping after 90 seconds if it doesn't.

03

Continue to work your fist down to the elbow, treating other points you find by repeating step 4.

Keep fist soft by imagining you are holding an egg

You could rest the arm on a table to keep it relaxed

BACK OF UPPER ARM

The muscle on the back of your arm is prone to overuse through a variety of activities, in particular when using crutches or a walking stick. Trigger points here cause pain that can be felt all the way from your neck and shoulder, to your hand and fingers. They can contribute to elbow pain and sensitivity, making it painful to rest your elbow on a table or chair.

01

To work this area, use an inflatable ball, or a trigger-point ball – whichever feels most comfortable.

POSSIBLE PAIN PATTERNS

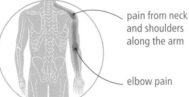

pain from neck and shoulders along the arm

elbow pain

MUSCLE LOCATOR

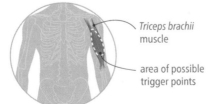

Triceps brachii muscle

area of possible trigger points

02

Use the fingers of your opposite hand to find any tender spots in the back of your upper arm.

05

Slowly move the ball to locate other tender spots, and repeat step 4 to ease the trigger points.

Make sure you work right to the outer edge of the back of the arm

Keep your arm relaxed

04

Hold this position until you feel a sense of ease, or for 90 seconds if the feeling doesn't change.

03

Place the ball on one of the tender spots and lean back against a wall.

02

Make a soft fist with your opposite hand and rest it on your forearm, just below the elbow. (If using a ball, place it between your forearm and the table.)

03

Use your knuckles to slowly "rake" down your forearm to locate trigger points. (If using a ball, gently press your arm into the ball and roll it down the arm.)

Palm rests
lightly on
the table

01

Rest your painful forearm palm down on a table or flat surface. (If you wish to use a ball, your arm should be palm-up.)

BACK OF FOREARM

The small extensor muscles in the back of your forearm straighten and extend your fingers. They work with your flexor muscles (see pages 84–85) to create the precise movements you need to use your fingers. These small muscles can tire quickly, creating trigger points. As well as pain, trigger points here can cause a sense of weakened grip.

04

When you find a tender spot, apply gentle but firm pressure until it eases, or for 90 seconds if it doesn't.

To make sure you don't press too hard, imagine you are holding an egg in your fist

05

Carry on moving either your fist or the ball down as far as your wrist. Stop at any tender points you find and repeat step 4.

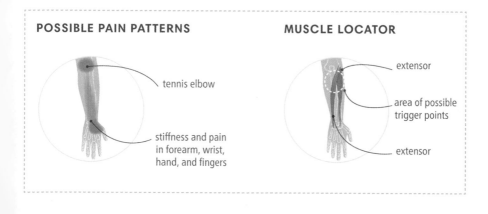

POSSIBLE PAIN PATTERNS

tennis elbow

stiffness and pain in forearm, wrist, hand, and fingers

MUSCLE LOCATOR

extensor

area of possible trigger points

extensor

02

Starting at the elbow, slowly roll your forearm over the ball. If using your hand, make a soft fist and slowly "rake" it down your forearm.

01

Put a ball on a table and rest your painful forearm on it, palm-down. If using your other hand instead of a ball, place your palm upwards.

FRONT OF FOREARM

The small muscles in the front of your forearm flex your fingers, working with the extensor muscles (see pages 82–83) to enable precise movements of the fingers. Like the extensors, these flexor muscles tire quickly and this can create trigger points. As well as arm and hand pain, trigger points here can cause your grip to become weaker.

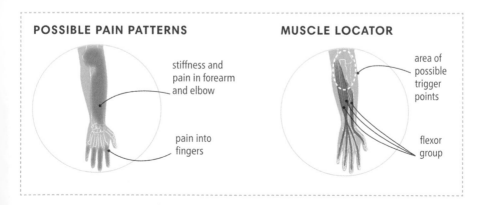

03

When you find a tender spot, stop and apply gentle but firm pressure until it eases, or for 90 seconds if it doesn't.

04

Carry on moving either the ball or your fist down to your wrist. Stop at any tender points and repeat step 3.

Use whichever type of ball feels most comfortable

POSSIBLE PAIN PATTERNS

stiffness and pain in forearm and elbow

pain into fingers

MUSCLE LOCATOR

area of possible trigger points

flexor group

PALM OF HAND

Your palms, fingers, and thumbs work together to execute fine movements such as writing, gardening, and chopping food. These small muscles tire easily with overuse, causing trigger points. Symptoms such as finger pain, numbness, stiffness, and pain on gripping items, are often misdiagnosed as repetitive strain injury, or what used to be called writer's cramp.

02

Choose a ball to work with – an inflatable ball for gentle, general pressure, or a smaller ball to pinpoint specific trigger points.

01

Find a table or other hard surface low enough that you can easily lean on it without raising your shoulder.

POSSIBLE PAIN PATTERNS

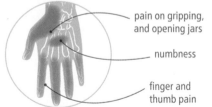

pain on gripping, and opening jars

numbness

finger and thumb pain

MUSCLE LOCATOR

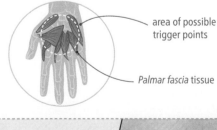

area of possible trigger points

Palmar fascia tissue

03

Place your hand on top of the ball and slowly and gently roll your whole hand, fingers, and wrist over it.

04

Stop at any tender areas you find and apply pressure until there is a feeling of ease, or stop after 90 seconds if there is no change.

05

Carry on moving your hand over the ball. Stop at any tender points you find and repeat step 4 to ease trigger points.

Keep arm relaxed

UPPER AND MID BACK

Back pain is the most common type of chronic pain in both the UK and the USA. Trigger points in your neck, shoulders, chest, and back can all contribute to pain and other symptoms felt in your upper and mid back.

HOW TRIGGER POINTS DEVELOP

• Too much time spent on computers and mobile devices locks your back into an unnatural position to support your arms and hands, which leads to trigger points.
• Slouching puts pressure on your back and causes more trigger points.
• Strain on your back muscles caused by incorrectly lifting and carrying heavy or awkward loads can lead to trigger points.

SYMPTOMS OF TRIGGER POINTS

• Sharp pains in and around your shoulder blades which are often worse when sitting still and ease as you move.
• Aching or pain in your mid back when you lift or carry heavy items. Often this leads people to say they have a weak back when trigger points may be the problem.

• A feeling that your shoulders are stuck in a rounded position or scrunched up around your ears. Stretching can help them loosen up but they soon go back to where they were.
• Tender spots in your back and sharp points of pain when you cough or breathe deeply. These usually come on after a cough or chest infection, or if you have asthma.
• Stiffness and pain in your back from spine conditions such as scoliosis (curvature of the spine) and ankylosing spondylitis (chronic inflammation of the spine). Working on trigger points can help to ease the discomfort from these conditions.

Pain patterns

The location and nature
of your back pain can help
you identify the best places
to look for trigger points
causing the pain.

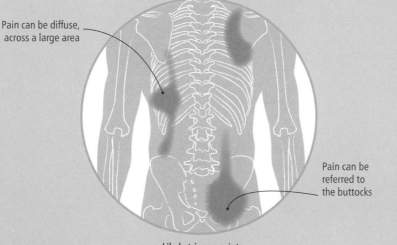

Pain can be
experienced
on either or both
sides of the body

Likely trigger points:
BETWEEN THE SHOULDER BLADES, pages 92–93

Pain can be diffuse,
across a large area

Pain can be
referred to
the buttocks

Likely trigger points:
BACK, pages 94–95

Continued ▶▶

Trigger-point locations

Use these body maps to
locate likely trigger points
and navigate to
pages that describe
how to treat them.

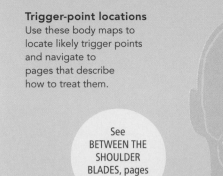

See
BACK OF
SHOULDER 2,
pages 58–59.

See
BETWEEN THE
SHOULDER
BLADES, pages
92–93.

See BACK,
pages 94–95.

See
SIDE OF BACK,
pages 108–109.

**BACK OF
BODY**

See BETWEEN
THE RIBS, pages
100–101.

See
SIDE OF CHEST,
pages 52–53.

**FRONT OF
BODY**

02

Put the balls on the floor
or on a bed, then lie on
them. If you prefer,
stand with your back to
a wall and lean against
the balls.

03

Allow your weight to
press down on the balls.
If you locate a tender
point, stop and let the
ball press against it.

01

To work this area, use
two tennis balls or soft
inflatable balls if you
prefer less pressure. Put
them in a pouch or bag.

BETWEEN THE
SHOULDER BLADES

The *Rhomboid* muscles attach the shoulder blades to the
spine and move your shoulders back when you pull something
towards you. If they get stuck in an overstretched position,
they cause rounded shoulders. Trigger points here cause an
achy pain on one or both sides of the upper back.

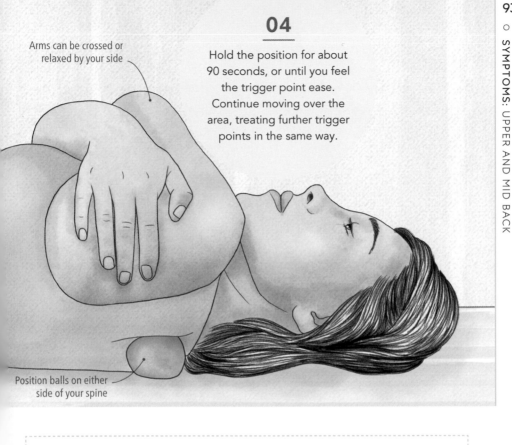

04

Hold the position for about 90 seconds, or until you feel the trigger point ease. Continue moving over the area, treating further trigger points in the same way.

Arms can be crossed or relaxed by your side

Position balls on either side of your spine

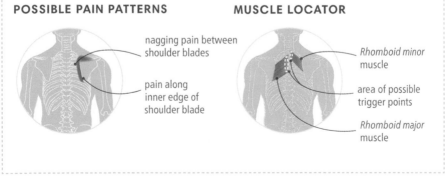

POSSIBLE PAIN PATTERNS

nagging pain between shoulder blades

pain along inner edge of shoulder blade

MUSCLE LOCATOR

Rhomboid minor muscle

area of possible trigger points

Rhomboid major muscle

BACK

Your back muscles are arranged in groups and layers, running from the pelvis to the head. Together they form the spine's support system, keeping your upper body stable and upright. Trigger points here cause pain, stiffness, and tightness, as well as sensitive or numb patches on the skin. As an alternative to the technique shown here, you could also use a therapy tool to work this area (see page 20).

POSSIBLE PAIN PATTERNS

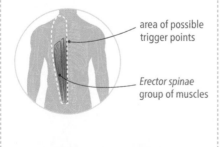

shoulder blade and neck pain

diffuse back pain that worsens through the day

pain in lower back and buttocks

MUSCLE LOCATOR

area of possible trigger points

Erector spinae group of muscles

01

Put two balls in a pouch or bag. Place the balls between your back and a wall. Alternatively, place the balls on the floor or a bed, and lie down on them, on your back.

02

Start at the top of your back. If you locate a tender point, let your weight press against it for 90 seconds, or until the tenderness eases.

03

Allow the balls to drop a little way down your back. Continue working any tender spots, holding for a maximum of 90 seconds on each trigger point you locate.

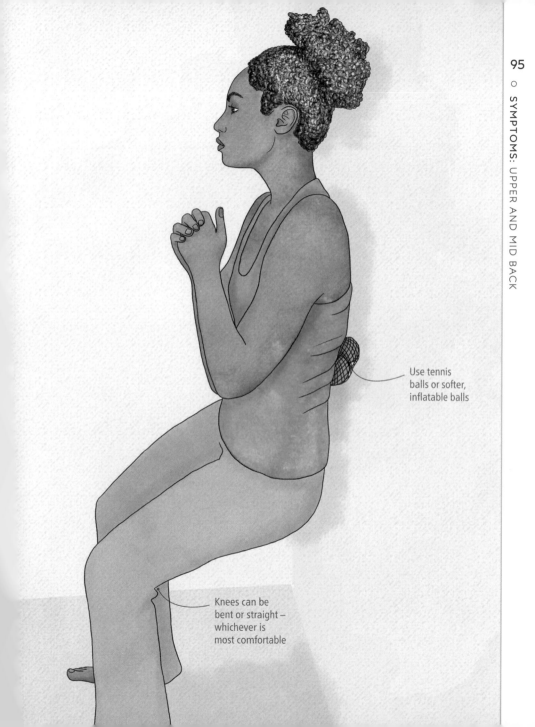

Use tennis balls or softer, inflatable balls

Knees can be bent or straight – whichever is most comfortable

ABDOMEN

Your abdomen contains essential organs such as the stomach, liver, bladder, and reproductive organs. Trigger points can develop in the muscles of your abdomen (your abs), as well as internally in the muscle tissue that lines the organs. If you experience unexpected abdominal discomfort or pain, consult a medical professional first, to rule out any other underlying causes.

HOW TRIGGER POINTS DEVELOP

• Strain on the abdominal muscles as a result of an infection, digestive upset, or cough can cause trigger points to develop in this area.

• Over-exertion of muscles, caused either by heavy lifting or repetitive actions such as sit-ups, can cause trigger points to form in the abdomen.

• Trigger points in other parts of the body, such as the muscles of your inner thighs and even your back muscles, can also cause pain in the abdominal area.

SYMPTOMS OF TRIGGER POINTS

• Trigger points in your organs and abdominal muscles can mimic the symptoms of medical conditions, such as: indigestion, heartburn, and nausea; bloating and colic; irritable bowel syndrome (IBS) and stomach upsets.

• Pain felt deep in the pelvis and period pain in women.

Pain patterns

The location and nature of your abdominal pain can indicate the most likely places to find the trigger points responsible.

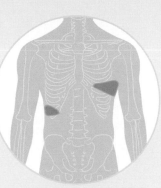

Likely trigger points:
BETWEEN THE RIBS, pages 100–101

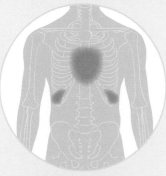

Likely trigger points:
DIAPHRAGM, pages 102–103

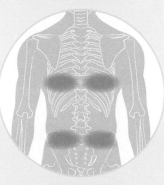

Likely trigger points:
ABDOMEN, pages 104–105

Continued ▶▶

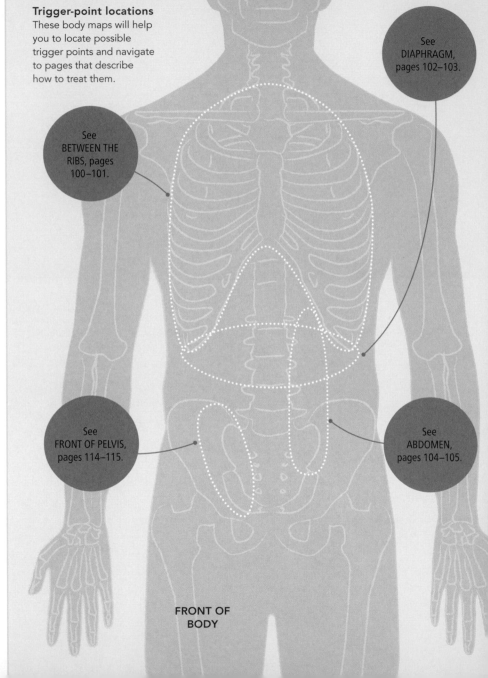

Trigger-point locations
These body maps will help
you to locate possible
trigger points and navigate
to pages that describe
how to treat them.

See
DIAPHRAGM,
pages 102–103.

See
BETWEEN THE
RIBS, pages
100–101.

See
FRONT OF PELVIS,
pages 114–115.

See
ABDOMEN,
pages 104–105.

**FRONT OF
BODY**

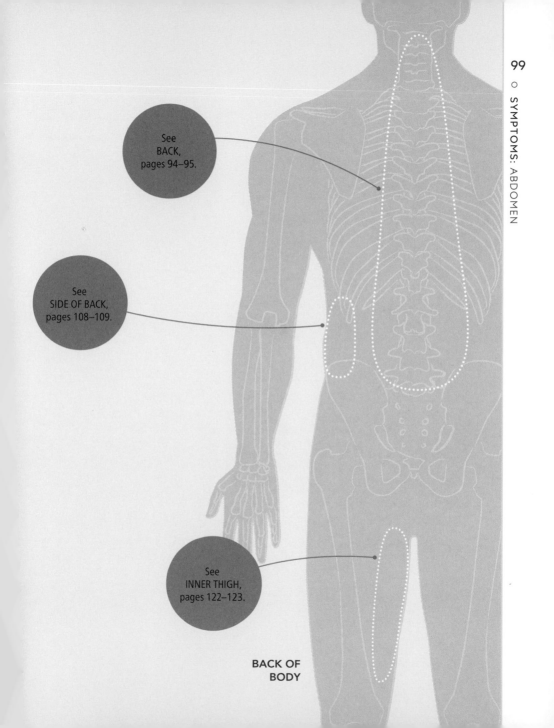

See
BACK,
pages 94–95.

See
SIDE OF BACK,
pages 108–109.

See
INNER THIGH,
pages 122–123.

**BACK OF
BODY**

BETWEEN THE RIBS

The intercostals are small muscles between the ribs that move the ribcage as you breathe. They can become stuck, especially if you have had a cold or a cough, and people with respiratory conditions such as asthma often develop trigger points in this area. If you have damaged or broken your ribs, don't work this area until they are fully healed.

03

Whenever you locate a tender spot, apply gentle but consistent pressure for about 90 seconds, or until the sensitivity eases.

02

Use your fingers to locate the small muscles between the ribs. Feel gently along the front and side of the ribs for areas that are tender when touched.

01

Stand in a comfortable position and bring the arm opposite your sore side across your body so your hand rests lightly on your ribcage.

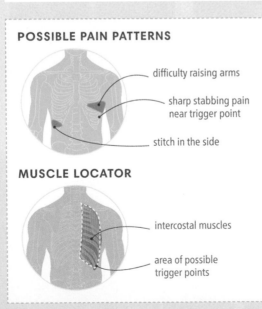

POSSIBLE PAIN PATTERNS

difficulty raising arms

sharp stabbing pain near trigger point

stitch in the side

MUSCLE LOCATOR

intercostal muscles

area of possible trigger points

04

To work the back of your ribs, use a small ball such as a trigger-point ball. Place it between a wall and your body, moving it over the back-rib area. Treat any trigger points you locate by repeating step 3.

DIAPHRAGM

The diaphragm is the main muscle used in breathing. It is a dome-shaped structure, tucked under the bottom of your ribcage. Together with the intercostals (see pages 100–101), the diaphragm allows your ribcage and lungs to expand and contract, enabling you to breathe. Trigger points can form here after a cold or cough, particularly in people who have respiratory conditions such as asthma.

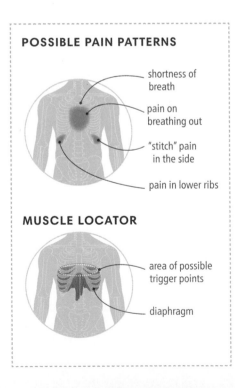

POSSIBLE PAIN PATTERNS

shortness of breath

pain on breathing out

"stitch" pain in the side

pain in lower ribs

MUSCLE LOCATOR

area of possible trigger points

diaphragm

01

Sit leaning slightly forward. Use your fingers to find the bottom of your ribcage. Curl your fingertips under your ribcage to press on the inside of the ribs.

02

Work from the centre of the ribcage towards your side. When you find a tender spot, press gently until the tenderness eases, for a maximum of 90 seconds.

03

Use the other hand to work from the centre of the ribcage out to your other side. Treat any trigger points by repeating step 2.

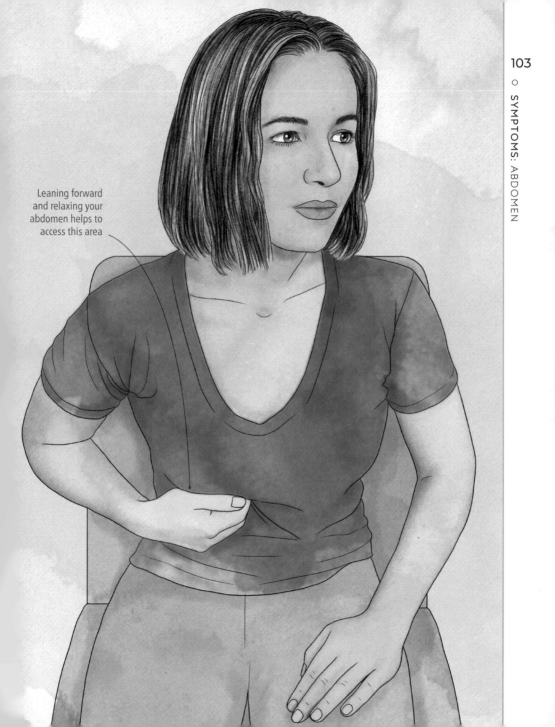

Leaning forward and relaxing your abdomen helps to access this area

ABDOMEN

The *Rectus abdominis* is your "six pack" muscle, and underneath it lie deeper muscles which wrap round your sides to connect with your lower back. Abdominals stabilize your body for a range of activities from breathing to sports, and are key muscles in childbirth. Trigger points can cause indigestion or nausea, and back pain which worsens with deep breathing, twisting or bending.

POSSIBLE PAIN PATTERNS

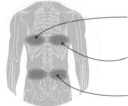

pain in mid-back area

heartburn or indigestion

lower back pain, including period pain

MUSCLE LOCATOR

Rectus abdominis muscle

area of possible trigger points

03

When you find a point that feels tender, use your finger pads to press firmly but gently into the area.

02

Use supported fingers by placing the fingers of one hand over the other to stroke gently over your whole abdominal area.

01

Sit on a bed, with your upper body well supported on pillows so that your abdomen is relaxed.

04

Maintain pressure until the pain eases, or stop after 90 seconds if it does not. Continue stroking the area and treat any more trigger points by repeating step 3.

Lean forward slightly to relax abdomen

Use finger pads to avoid digging into skin

LOWER BACK AND HIPS

Long-term lower back pain is debilitating and can affect both physical and mental health. Trigger points in the muscles of the back, buttocks, sides of the hips, and groin contribute significantly to chronic pain in this area.

HOW TRIGGER POINTS DEVELOP

• Strains and sprains from lifting heavy objects can cause trigger points.
• Switching from being sedentary to doing strenuous exercise, without the underlying core strength to support your body, can also result in trigger points.

SYMPTOMS OF TRIGGER POINTS

• Sciatica – pain and numbness that can travel from your buttock to your foot.
• Lower-back pain and stiffness, aggravated by twisting or lifting.
• Piriformis syndrome, which affects the muscles found deep within the buttocks..
• Chronic pelvic pain syndrome – the umbrella term given to persistent pain in the area around hips and pelvis.

PAIN PATTERNS

The location and nature of your lower hip and back pain can indicate likely places to look for the trigger points responsible.

Likely trigger points:
SIDE OF BACK,
pages 108–109

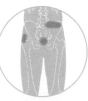

Likely trigger points:
BUTTOCKS,
pages 110–111

Likely trigger points:
SIDE OF HIP 1,
pages 112–113

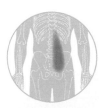

Likely trigger points:
FRONT OF PELVIS,
pages 114–115

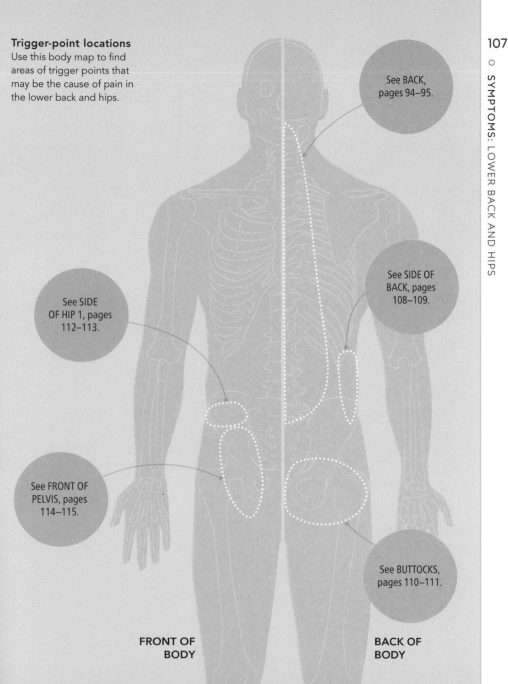

Trigger-point locations
Use this body map to find
areas of trigger points that
may be the cause of pain in
the lower back and hips.

See BACK,
pages 94–95.

See SIDE OF
BACK, pages
108–109.

See SIDE
OF HIP 1, pages
112–113.

See FRONT OF
PELVIS, pages
114–115.

See BUTTOCKS,
pages 110–111.

FRONT OF
BODY

BACK OF
BODY

SIDE OF BACK

The *Quadratus lumborum* is a deep muscle that connects the bottom of your ribs to the top of your pelvis, and is also attached to the side of your spine in your lower back. It stabilizes your body and allows it to bend to the side. Trigger points here cause pain in various zones including the lower abdomen, the sacroiliac (SI) joints at the back of your pelvis, and the groin.

POSSIBLE PAIN PATTERNS

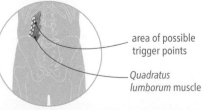

sharp pain
on coughing
or sneezing

hip pain

buttock pain

MUSCLE & TRIGGER POINT LOCATOR

area of possible
trigger points

*Quadratus
lumborum* muscle

03

Slowly move your body over the ball. When you locate a tender point, lean into the ball to apply gentle pressure to the spot.

02

Place an inflatable ball on the area and lean against a wall. If you prefer, you can lie down with the ball between your back and the bed.

01

Feel your sore side with the fingers of your opposite hand to find the area of soft tissue between the bony structures of your ribs and pelvis.

04

Move on when the tenderness eases, or after 90 seconds if it does not. Treat any further trigger points you find in the same way.

05

To work trigger points that may be tucked under your ribs or at the top of your pelvis, it may help to use a specialist tool (see page 20).

Position your arm in the most comfortble way

BUTTOCKS

Your buttocks are formed by very strong muscles that enable you to walk, run, climb stairs, or stand up from sitting in a chair. Trigger points here cause pain in the buttocks when performing any of these activities, as well as pain if you sit in the same position for any length of time. Trigger points in the *Piriformis* muscle can cause sciatica.

POSSIBLE PAIN PATTERNS

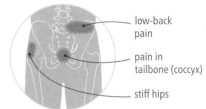

— low-back pain

— pain in tailbone (coccyx)

— stiff hips

MUSCLE LOCATOR

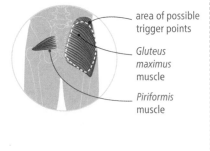

— area of possible trigger points

— *Gluteus maximus* muscle

— *Piriformis* muscle

01

Use a soft, inflatable ball for this exercise, as it provides a wider area of pressure.

02

Sit on the ball, either on the floor with a wall to support your back, or on a firm chair.

03

If you feel a tender point, allow the weight of your body to sink into the ball and wait until the sensation eases.

04

Stop after 90 seconds if the tenderness does not ease. Treat other trigger points in the area in the same way.

If you feel a sharp pain, move the ball slightly, away from the sciatic nerve

SIDE OF HIP 1

The muscles in the side of your hip rotate your leg within the hip socket, which is key to activities such as walking and running. Trigger points in this area can hamper or prevent this movement; as well as causing pain, they can make it uncomfortable to sleep on your side and walking or running becomes almost impossible.

03

Place an inflatable ball on the muscle and lean on it against a wall. Alternatively, lie down either on the floor or on a bed. Use your body to move the ball slowly over the area.

02

Feel about 5cm below this point for the *Gluteus medius* muscle, which runs in a line from here along the top of your buttock.

01

To locate the muscle, feel for the top of your hip bone and follow it round to the side of your hip.

POSSIBLE PAIN PATTERNS

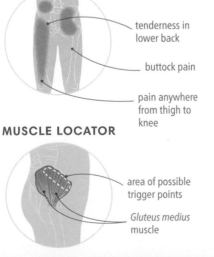

tenderness in lower back

buttock pain

pain anywhere from thigh to knee

MUSCLE LOCATOR

area of possible trigger points

Gluteus medius muscle

04

When you locate a tender spot, maintain pressure until it feels easier, or for 90 seconds. Locate and work other trigger points by repeating steps 3–4.

Trigger points in this area can feel especially intense

01

To find the right area to work, feel on one side of your lower abdomen with your fingers to locate your pelvic bone.

FRONT OF PELVIS

The *Iliopsoas* is a deep muscle that moves your hip when you walk, run, or go up stairs and hills. It can develop tightness and trigger points from either too much or too little use. Trigger points here cause back pain, with difficulty sitting up or standing for long periods. They can also affect breathing if they restrict the movement of your diaphragm.

02

Place an inflatable ball just to the inside of the bone on the soft tissue, then lie on your front on the floor or on a bed.

03

As the ball pushes into tender areas, you may feel indigestion-like discomfort. Maintain gentle pressure until the tenderness eases, stopping after 90 seconds if it does not.

04

Now move the ball to work the other side of your pelvis. Repeat step 3 to work other trigger points that you locate.

POSSIBLE PAIN PATTERNS

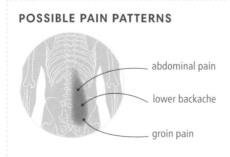

abdominal pain

lower backache

groin pain

MUSCLE LOCATOR

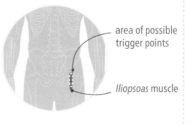

area of possible trigger points

Iliopsoas muscle

LEGS AND KNEES

Our legs have to support our weight and withstand the impact of us walking and running. The knees are among the hardest-working of all the body's joints, so it is not surprising that one-fifth of all adults suffer knee pain at some time, and that the likelihood of problems increases with age.

HOW TRIGGER POINTS DEVELOP

• Twisting your knee can cause ligament and muscle damage, weakening your knee and leading to trigger points.
• Trigger points can be caused by muscle strains and sprains, or by soft-tissue damage following a fall.
• Trigger points can develop in your hips and legs if you participate in sports where you use your legs repetitively, for example, running, cycling, or football.
• Trigger points in the muscles of the hips, buttocks, groin, and thighs can refer pain and other symptoms into the legs and knees.
• Muscle strains in the hamstring and thigh muscles can create trigger points that leave the muscles vulnerable to further painful strains.

SYMPTOMS OF TRIGGER POINTS

• Knee pain and stiffness, generally caused by tightening of the tissues supporting the joint. This often follows ligament damage which causes short-term instability.
• Patello-femoral syndrome is pain felt under the kneecap especially when walking up or down hills or stairs. This can be caused by wear and tear in the joint, but is often due to trigger points in your thigh muscles.
• Shin splints, which are sharp pains in your shins that come on when you run, are typically caused by trigger points.
• Calf problems, including an aching, burning, or cramping pain; numbness and weakness in the area; and cramps and restless legs at night.

Pain patterns

The location and nature of your leg and knee pain can help identify likely places to look for the trigger points responsible.

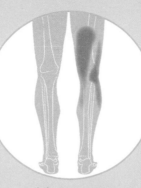

Likely trigger points:
BACK OF THIGH, pages 120–121

Likely trigger points:
INNER THIGH, pages 122–123

Likely trigger points:
FRONT OF THIGH, pages 124–125

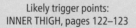

Likely trigger points:
SIDE OF HIP 2, pages 126–127

Likely trigger points:
CALF, pages 128–129

Continued ▶▶

Trigger-point locations
Use these body maps to
locate trigger points and
navigate to the pages that
describe how to work them.

See
FRONT OF PELVIS,
pages 114–115.

See
INNER THIGH,
pages 122–123.

See
FRONT OF THIGH,
pages 124–125.

See
SIDE OF CALF,
pages 134–135.

**FRONT OF
BODY**

See
SHIN,
pages 132–133.

See
SIDE OF HIP 1,
pages 112–113.

See
SIDE OF HIP 2,
pages 126–127.

See
BACK OF THIGH,
pages 120–121.

See
CALF, pages
128–129.

**BACK OF
BODY**

BACK OF THIGH

This area is made up of a group of three muscles called the hamstrings. Tight hamstrings cause pain and, as they are attached to your pelvis, they can also pull on the bones, tilting your pelvis back and causing lower-back pain and difficulty sitting cross-legged. Trigger points here can cause pain in the back of your thigh when sitting or walking, sometimes making you limp.

02

Place the ball under your thigh, just under the crease of your buttock, and let the weight of your leg rest on the ball.

01

To work on the hamstring area, it's best to use an inflatable ball. Sit on the floor and lean against a wall, or sit on a chair.

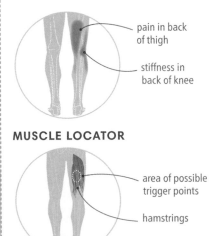

POSSIBLE PAIN PATTERNS

— pain in back of thigh

— stiffness in back of knee

MUSCLE LOCATOR

— area of possible trigger points

— hamstrings

03

Allow the ball to press into the area until you feel the tissues soften, stopping after 90 seconds if they don't.

04

Move the ball slowly over the whole thigh area, from the top to just above the knee. Stop at each tender spot you find, and repeat step 3 to ease trigger points.

Avoid the back of the knee, where there are sensitive nerves and blood vessels

INNER THIGH

Your inner thigh is formed of a group of five muscles known as the adductors. They work to bring your legs together, and are the cause of groin strain, especially in athletes and dancers. Trigger points here can cause pain and stiffness in your thigh when you open or rotate your leg outwards. They can also cause clicky hips and pain felt deep inside your pelvis or hip joint, which is often attributed to other causes such as a hernia or arthritis.

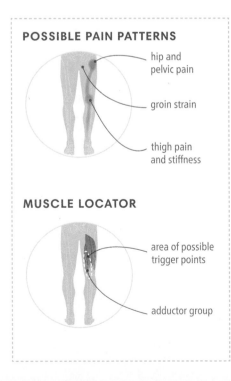

POSSIBLE PAIN PATTERNS

hip and
pelvic pain

groin strain

thigh pain
and stiffness

MUSCLE LOCATOR

area of possible
trigger points

adductor group

01

Sit on the edge of a hard chair and place an inflatable ball under the top of your inner thigh. If you prefer, you can lie on your sore side on the floor with your other leg bent out of the way.

02

Rest your leg on the ball. If you feel a tender spot, allow the weight of your leg to press the ball into the area for 90 seconds, or until the tissues start to ease.

03

Move the ball slightly and repeat step 2 as you locate any further tender or sensitive spots.

Use an inflatable ball for this sensitive area

01

Lie on the floor or your bed and place the ball under the front or side of your thigh. You can also work on this area by leaning on the ball against a wall.

02

Allow your leg to relax onto the ball. If you feel tenderness, maintain the pressure for 90 seconds, or until you feel the tissues ease.

An inflatable ball works a wide area, while a trigger-point ball is more focused

FRONT OF THIGH

The quadriceps are a group of four muscles that form the front of your thigh, passing over your knee joint. Trigger points here can cause knee pain, particularly under the kneecap. They can make it hard to bend your knee, or can make your knee give way. Trigger-point pain here can be mistaken for tendonitis, bursitis, or arthritis of the knee.

03

Move the ball slowly over the area, and repeat step 2 to work any further trigger points you find.

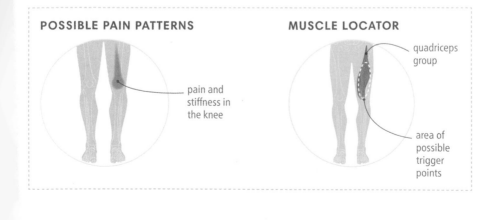

POSSIBLE PAIN PATTERNS

pain and stiffness in the knee

MUSCLE LOCATOR

quadriceps group

area of possible trigger points

SIDE OF HIP 2

The *Tensor fasciae latae* (TFL) is a muscle at the top of your hip that connects with the iliotibial band (ITB), a tendon that runs down the outside of the leg and over the knee. Trigger points can cause hip and knee pain, and when the TFL muscle is tight it can contribute to runner's knee or ITB syndrome. You may feel pain when lying on your side, walking, or running, although these symptoms can also be caused by the quadriceps muscles in the front of your thigh (see pages 124–125).

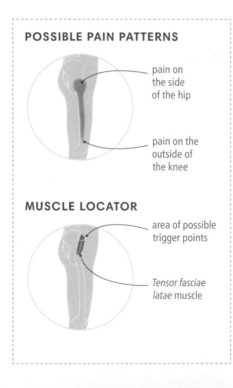

POSSIBLE PAIN PATTERNS

pain on the side of the hip

pain on the outside of the knee

MUSCLE LOCATOR

area of possible trigger points

Tensor fasciae latae muscle

01

Locate the TFL by using your fingers to find the bony point at the front of your hip. Move back from this point to find the muscle.

02

Place an inflatable ball on the area and lean against a wall. If you prefer, lie down with the ball between you and the floor.

03

Lean into the ball. If the spot feels tender, apply pressure for 90 seconds, or until the feeling eases. Move the ball over the area, applying pressure to any sensitive points you find.

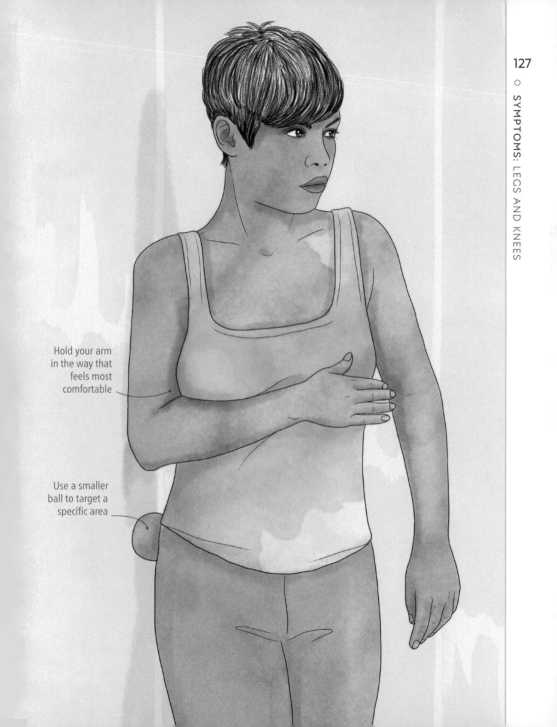

Hold your arm in the way that feels most comfortable

Use a smaller ball to target a specific area

CALF

The calves are your body's main weight-bearing area. They can become tight from a variety of activities, such as running, cycling, wearing high heels or simply from sitting for long periods. As well as causing calf problems, trigger points here are also a major cause of foot pain, such as plantar fasciitis. They can also contribute to lower-back pain.

02

Sit on the floor with your back to a wall, or sit along a sofa. Place the ball under the top part of your calf and let your leg rest on it.

01

Use an inflatable ball to work the area. For more intense pressure, use a trigger-point ball.

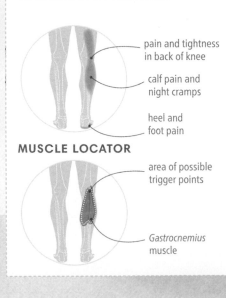

POSSIBLE PAIN PATTERNS

pain and tightness in back of knee

calf pain and night cramps

heel and foot pain

MUSCLE LOCATOR

area of possible trigger points

Gastrocnemius muscle

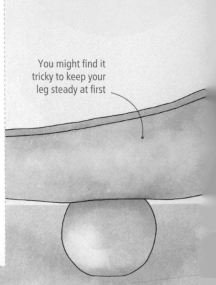

You might find it tricky to keep your leg steady at first

03

Allow the weight of your leg to press the ball into your calf for 90 seconds, or until you feel the tissues under the ball softening.

04

Move the ball slowly down your calf to find any other trigger points, stopping and repeating step 3 to treat them.

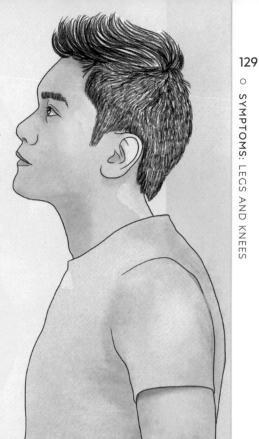

ANKLES
AND FEET

Our feet act both as springs pushing the legs off from the ground, and shock-absorbers when they land. These complex, weight-bearing structures are vulnerable to pain from trigger points in the muscles and soft tissues of our lower legs, ankles, and feet.

HOW TRIGGER POINTS DEVELOP

• Overuse of the muscles, caused by walking, running, or standing for long periods can cause trigger points.
• Wearing high heels or tight shoes can restrict the muscles.
• Trigger points in the upper legs and hips can refer pain to the feet.
• Sprains can weaken the ankle ligaments. This in turn can cause trigger points that lead to more sprains.
• Trigger points in your lower leg can cause pain and swelling in the ankle.

SYMPTOMS OF TRIGGER POINTS

• Plantar fasciitis is an inflammation of the sole of the foot, which makes it painful to stand or walk. Symptoms tend to be worse in the morning, and can include sensations of cramp, although these ease with walking.
• Policeman's heel is a similar condition to plantar fasciitis, with the heel becoming sore, instead of the sole of the foot. It is most common among people who spend a lot of time on their feet.

PAIN PATTERNS

The location and nature of your pain can indicate likely places to look for the trigger points responsible.

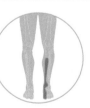

Likely trigger points:
SHIN, pages 132–133

Likely trigger points:
SIDE OF CALF,
pages 134–135

Likely trigger points:
SOLE OF FOOT, pages
136–137

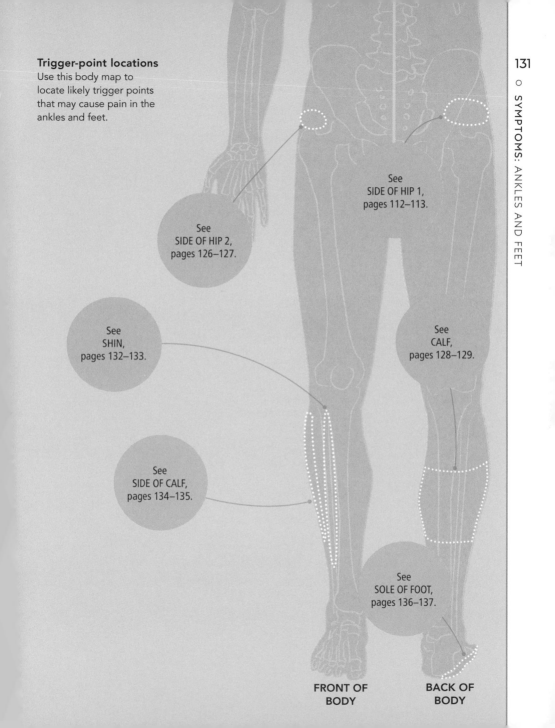

Trigger-point locations
Use this body map to
locate likely trigger points
that may cause pain in the
ankles and feet.

See
SIDE OF HIP 1,
pages 112–113.

See
SIDE OF HIP 2,
pages 126–127.

See
SHIN,
pages 132–133.

See
CALF,
pages 128–129.

See
SIDE OF CALF,
pages 134–135.

See
SOLE OF FOOT,
pages 136–137.

FRONT OF
BODY

BACK OF
BODY

02

Lie on your side,
either on the floor or
on a bed, with your
sore leg bent.

03

Place a ball at the top of
the muscle and let your
leg rest on the ball for
90 seconds, or until you
feel the tissues start
to soften.

01

Locate the shin muscle
by feeling along the
outside of your shin.
If you lift your foot up
you will see the outline
of the muscle.

Use either an inflatable
ball or a trigger-point
ball, whichever feels
most appropriate

SHIN

Your shin muscle runs down the outside of your lower leg and wraps over
the top of your foot into the arch. It's attached to the bone all the way
down your shin. When the muscle is too tight it can pull against the bone,
causing pain known as shin splints. Trigger points here can cause pain in
your ankle or big toe (sometimes misdiagnosed as gout). You might also
find it painful to lift your foot when walking or running.

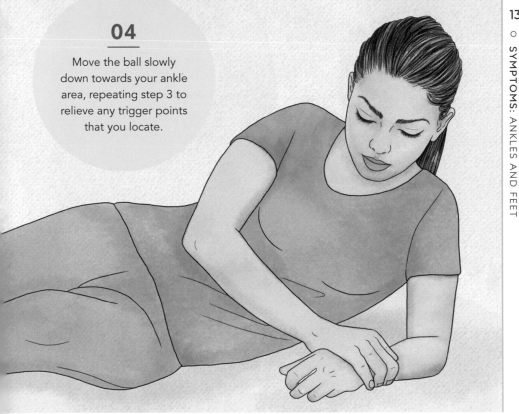

04

Move the ball slowly down towards your ankle area, repeating step 3 to relieve any trigger points that you locate.

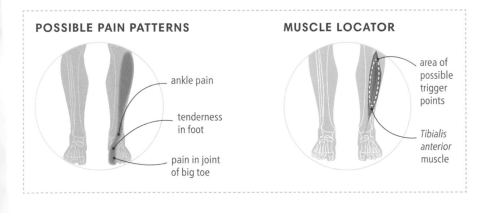

POSSIBLE PAIN PATTERNS

ankle pain

tenderness in foot

pain in joint of big toe

MUSCLE LOCATOR

area of possible trigger points

Tibialis anterior muscle

SIDE OF CALF

This muscle group follows the line of a trouser seam down the outside of your lower leg to your ankle, and then wraps under your foot. Trigger points here can cause pain and stiffness in the outside of your ankle and the side of your foot. They can also mimic the symptoms of a sprained ankle, but without any swelling, and can make your ankle feel weak.

01

Find this muscle group by placing your fingers on the outside of your lower leg. When you turn your foot outwards, you'll feel the muscles contract.

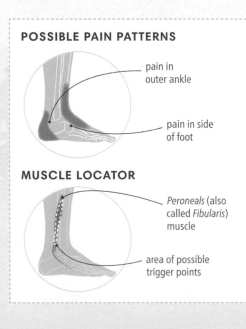

POSSIBLE PAIN PATTERNS

pain in outer ankle

pain in side of foot

MUSCLE LOCATOR

Peroneals (also called *Fibularis*) muscle

area of possible trigger points

02

Lie on your painful side on the floor or on your bed, with your lower leg bent slightly.

03

Place a ball at the top of the area you located. Allow the weight of your leg to press the ball gently into the muscle.

04

Wait for a sense of the tissues softening, or for a maximum of 90 seconds. Gradually move the ball down towards the ankle, treating any further tender areas by repeating step 3.

SOLE OF FOOT

Your feet are your base and your whole weight rests on them – pain here can affect many other parts of your body. Trigger points can develop in the soles of your feet from everyday activities such as walking or standing, or from wearing high heels. These trigger points can cause foot and heel pain, soreness in the sole of your foot (also called plantar fasciitis), stiffness, and numbness. They can also contribute to problems in your ankles, knees, and hips.

POSSIBLE PAIN PATTERNS

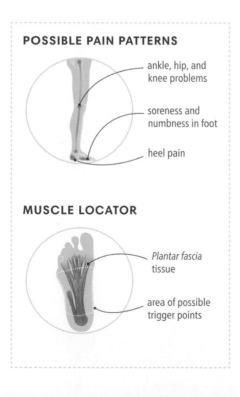

ankle, hip, and knee problems

soreness and numbness in foot

heel pain

MUSCLE LOCATOR

Plantar fascia tissue

area of possible trigger points

01

Stand or sit, whichever feels most comfortable, in your bare feet.

02

Place an inflatable ball under your sore foot. Gently and slowly roll your foot over the ball.

03

If you feel a tender spot, stop and let your weight push the ball against it for 90 seconds, or until the feeling eases.

04

Continue to roll the ball over the foot and treat further tender points by repeating step 3.

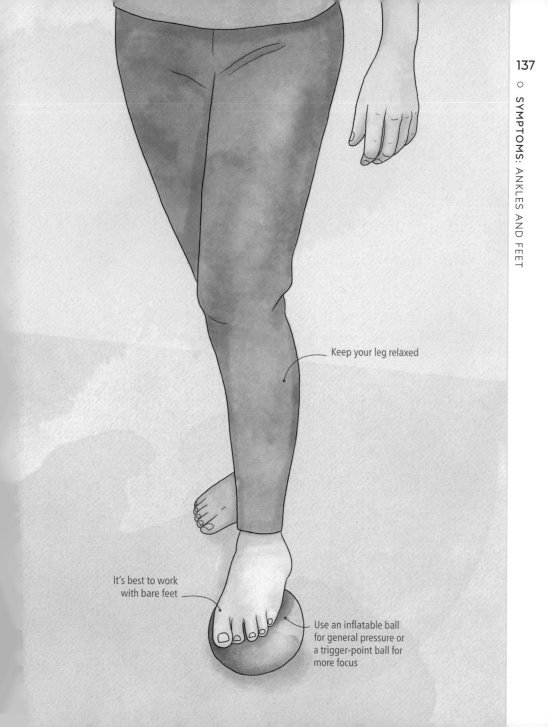

Keep your leg relaxed

It's best to work
with bare feet

Use an inflatable ball
for general pressure or
a trigger-point ball for
more focus

RESOURCES

An understanding of trigger points goes hand-in-hand with an understanding of fascia – the main connective tissue in the body. There are many resources aimed at the therapist. For those readers who would like to know more but without the in-depth anatomy and physiology training of the professionals, here are some useful suggestions:

FURTHER READING

The Trigger Point Therapy Workbook
by Clair Davies (New Harbinger Publications, 2004)
An idiosyncratic in-depth guide to self-treatment for pain relief.

Living Pain Free: Healing Chronic Pain with Myofascial Release
by Amanda Oswald (Lotus Publishing, 2017)
A self-help guide that explains clearly and simply the relationship between fascia and chronic pain. Includes simple self-care techniques, stretches and exercises which are a perfect complement to this book. Available as a book, Kindle edition and online video programme.

ONLINE

Jean-Claude Guimberteau
A French hand surgeon who has specialised in filming fascia in the living body. His fascinating videos help show how the body is interconnected and how restrictions and trigger points in one area can affect other parts. Extracts are widely available on YouTube and can be accessed by typing in his name.

EQUIPMENT

The inflatable and trigger point balls featured in this book are available as a Myofascial Release Kit from Pain Care Clinic – paincareclinic.co.uk – and Amazon.

Other trigger point tools are available from Amazon and other online retailers.

THERAPISTS

Trigger-point therapy is available from many massage and other bodywork therapists. Trigger-point therapy from a therapist who has an understanding of fascia and myofascial release offers an holistic approach to your pain condition.

paincareclinic.co.uk – UK-based clinics run by Amanda Oswald and specialising in myofascial release and trigger-point therapy.

myofascialtherapy.org – US-based directory of myofascial trigger-point therapists which includes listings for UK therapists.

mfrtherapists.com – US-based directory of myofascial therapists which includes listings for other countries including the UK.

myofascialrelease.uk – UK-based directory of myofascial therapists which includes listings for some other countries.

INDEX

Index entries in **bold** indicate specific trigger-point treatments

○

ABOUT THE AUTHOR

Amanda Oswald is a leading UK myofascial-release and trigger-point specialist who has spent the last 11 years working with people who have complex chronic pain conditions. She runs the Pain Care Clinic with facilities in several UK locations including London's prestigious Harley Street. Using advanced bodywork techniques and cognitive hypnotherapy, she takes a mind-body approach to healing chronic pain, particularly encouraging people to empower and help themselves out of pain using simple myofascial and trigger-point techniques. She is author of the self-help guide *Living Pain Free: Healing Chronic Pain with Myofascial Release.* In addition to her clinic work, Amanda runs workshops teaching the public self-help for chronic pain. She has also developed the Practical Myofascial Release qualification for therapists.

AUTHOR'S ACKNOWLEDGMENTS

I would like to thank all those who have helped me on my journey into a better understanding of chronic pain and how to treat it. In particular I would like to thank my clients. Each and every one of them has taught me something more about the connections between pain and the fascinating world of fascia and trigger points. My teachers have been many over the years and to them I would like to say thank you – you have helped to inspire me and to weave a deep web of knowledge. Particular thanks go to Kahn Priestley who has continued to provide unfailing support and indulged my tendency to go off at tangents.

PUBLISHER'S ACKNOWLEDGMENTS

DK would like to thank the following people for their assistance in the publication of this book: Mandy Earey for additional illustrations and design work, Steve Crozier for retouching, John Friend for proofreading, and Marie Lorimer for compiling the index.